MODERN HYDRAULICS
The Basics at Work

William Wolansky
Arthur Akers

Iowa State University of Science and Technology

Amalgam Publishing Company

P.O. Box 7138
San Diego, California 92107

MERRILL PUBLISHING COMPANY
A Bell & Howell Information Company
Columbus   Toronto   London   Melbourne

To our students, who stimulate us and teach us so much.

Cover Photo: Courtesy Chrysler Corporation

Published by Merrill Publishing Company
A Bell & Howell Information Company
Columbus, Ohio 43216

This book was set in Century Schoolbook

Administrative Editor: John Yarley
Art Coordinator: Pat Welch
Cover Designer: Cathy Watterson

Library of Congress Catalog Card Number: 87-62818
International Standard Book Number: 0-675-20987-0
Printed in the United States of America
1 2 3 4 5 6 7 8 9—92 91 90 89 88

MERRILL'S SERIES IN MECHANICAL AND CIVIL TECHNOLOGY

FOREWORD

An effective educational program is usually heavily dependent upon accurate, clear, concise, and timely text materials. Having written much text material, I can appreciate the authors' work in choosing material and writing it at the appropriate academic level. The authors have tried to bridge the communication gap between the graduate and the work community.

This text assists the learning process with appropriate mathematics as support and guidance for practical hands-on use of fluid power training.

Students who wish to explore subjects beyond the range of this text can consult the references to current specialized fluid power educational materials presented at the ends of the chapters.

As Educational Coordinator for the Fluid Power Educational Foundation, I am pleased to see *MODERN HYDRAULICS: THE BASICS AT WORK* available to the educational community. I extend my personal congratulations and those of the Foundation Trustees for this contribution to fluid power education.

John J. Pippenger, P.E.
Educational Coordinator
Fluid Power Educational Foundation

PREFACE

This teaching text introduces and reinforces key principles, concepts, and related component information that help build a foundation for career work in fluid power technology. While books on this subject are available, none of them places sufficient emphasis on the practical applications of theory. This book is dedicated to fill that void. Because it is written for industrial technology students, technicians, and service maintenance personnel, it does not teach advanced fluid power topics. Scientific principles are explained in simple, illustrated forms and patiently reinforced with practical, easy-to-grasp exercises. This text enables a student to grasp fluid power principles, see their applications at work in a variety of industries and types of equipment, and finish with a solid understanding of the "why" and "how" of various operating fluid power components and systems.

We use a systems approach to assure comprehension of key concepts and the relationships of fluid power behavior within a dynamic operational system throughout the subsystems and, ultimately, the components. The algebra and physics that are used are reinforced by application-oriented problems and suggested activities. Suggested readings at the end of each chapter point the reader to additional sources. This textbook features building-block chapters essential to understanding theory, fluid conditioning, components, circuit design and operation, and safety.

Chapter 1 is an introduction to fluid power technology. The reader gets an overview of the many applications of fluid power, the historical development of fluid power technology, and the advantages of fluid power for transmitting and controlling power. This foundation chapter creates an awareness of the diversity of applications of fluid power systems as they were designed to do useful work.

Chapter 2 discusses fundamental principles and applications of fluid mechanics to demonstrate effective use of static fluids to perform useful work. Chapter 3 builds the laws of fluids in motion and provides concepts of continuity momentum and energy. Applications of these concepts to reinforce dynamic fluid power principles are then illustrated. Chapter 4 includes a discussion of accessories in a hydraulic system such as reservoirs, conductors, and fluid conditioners. It covers the design, maintenance, and safety of operating fluids within a hydraulic system. Confining a fluid and conducting it from the reservoir to the control and work elements of a hydraulic system are important to the safety and efficiency of the operational system, as well as to the life wear of the components within the system. Chapter 5 describes the different types of fluids available for use in hydraulic systems and outlines the importance of their cleanliness, how cleanliness is achieved, and what we might expect in future fluid developments. Chapter 6 illustrates how pumps can create flow of fluids in enclosed hydraulic application circuits and can be used efficiently to transmit power. Chapter 7 describes how the flow of fluid can be controlled in terms of pressure, flow, and sequence to accomplish work with appropriate control devices. Chapter 8 discusses the use of actuators to perform work safely and efficiently. Chapter 9 outlines the elements in the design of a fluid power circuit and stresses safety as an important consideration when working with fluid power devices or systems. Chapter 10 includes a discussion of how the development of fluid power control will be affected by the technological advances currently used in electrical and electronic interfaces. Chapter 11 discusses pneumatics. The principles are developed by studying the properties of air, but the same principles could apply to almost any gas.

The step-by-step approach of this text provides a logical sequence and interface of theory, application, and practical design, assuring a solid understanding of basic hydraulics. After studying this text, the student will be familiar with terms, symbols, equations, hardware components, and system applications of basic hydraulics.

This book is a basic text that provides theory, component information, circuit design, and student problems. It also includes laboratory experiences for students to verify and record what actually happens in a dynamic circuit as contrasted with theoretical computations.

A successful mastery of hydraulics requires that a balance of time be spent learning the theory and applying that knowledge in practice. Resourceful teachers will use films, slides, transparencies, and cut-away components as well as trainers to enable students to learn, understand, and apply the content of this text.

The reader should become familiar with the expectations of each chapter and the new terms introduced early in the chapter. After reading each chapter, the student should be able to answer the questions and problems at the end of the chapter. For more advanced discussion of topics, suggested reading sources are listed at the ends of chapters.

We take this opportunity to express our appreciation for the efforts of all who brought this product to fruition. We gratefully acknowledge the advice provided by the prepublication reviewers: Victor Bridges, Umpqua Community College; Kurt Keydell, Montgomery College; Nikhil Kundu, North Texas State University; and Leonard T. Turn, Pennsylvania Institute of Technology. The generosity of the many organizations and fluid power component manufacturers who gave us permission to use illustrations and photographs is particularly appreciated. Our illustrator, Lynn Ekblad, also deserves praise. To all others who helped in any way, we express our appreciation.

We have left until last the pleasure of thanking our wives, Claire Wolansky and Marcia Akers. Without their care and patience, this work would not have been possible.

CONTENTS

CHAPTER 1

Development of Fluid Power Technology

Introduction

In this chapter, the definition and development of fluid power technology, showing the diversified applications and advantages of fluid power systems, and the suggested learning activities will provide an overview of fluid power technology. After reading this chapter, you will be able to define fluid power, trace the historical development of fluid power, name the advantages of hydraulic systems and their applications, and answer the end-of-chapter questions.

Key Terms

Fluid Power: The transfer of power by means of potential energy changes in a fluid flowing at a controlled rate.

Hydraulics: Engineering science pertaining to liquid pressure and flow.

Pneumatics: Engineering science pertaining to air pressure and flow.

Power: The rate of doing work, or the rate of energy transfer. In conventional units (CU): 1 horsepower (hp) = 550 ft · lb/s = 33,000 ft · lb/min. In SI units: P = joules/s = watts.

Pressure: Force per unit area. The distributed reaction (pressure) on a confined fluid is measured in pounds per square inch (psi) or in SI units, pascal (Pa) = N/m^2.

Work: The product of the applied force and the distance an object moves. In CU units: lb × ft = ft · lb, or in inches = in. · lb. In SI units: N · m.

1.1 WHAT IS FLUID POWER?

Fluid power technology is a means to convert, transmit with control, and apply fluid energy to perform useful work. Since a fluid can be either a liquid or a gas, *fluid power* is the general term that includes hydraulics and pneumatics. Fluid power specifically deals with the transfer of power by means of potential energy changes in a fluid flowing at a controlled rate. The others are mechanical (shafts, gears, pulleys, clutches), or electrical, such as generators or motors.

Liquids provide a very suitable medium for transmitting power. They can be controlled to ensure the work accomplished by a cylinder or motor is done smoothly, accurately, efficiently, and safely.

Fluid power is versatile: an electrical solenoid may switch a control valve that directs fluid to an actuator, which exerts several tons of force on the shovel of a front-end loader; or, it may control with preciseness and repetition the machining of components in a manufacturing process.

1.2 THE HISTORY OF FLUID POWER

The use of moving fluids to perform work dates back as far as 5000 years. The Chinese and Egyptians used wind and moving water to do useful work. Early Chinese records indicate wooden valves were used to control water flow through bamboo pipes about 4000 B.C. Ancient Egyptians built a masonry dam across the Nile, 14 miles south of present-day Cairo, for control of irrigation water by canals, sluices, brick conduits, and ceramic pipes. During the Roman Empire, extensive water systems using aqueducts, reservoirs, and valves were constructed to carry water to cities.

Historically, the use of water to drive water wheels and air to propel windmills or sail ships depended on the movement of vast quantities of fluid at relatively low pressure. People also relied upon nature to supply the pressure, such as that created by wind or a waterfall. Egyptians quarried marble for large structures by drilling holes into the marble, adding water to the holes, and then compressing the water by driving in wooden plugs.

In spite of the early uses of water and air for doing work, the principles of fluid flow were not well understood until 1650, when the French scientist Blaise Pascal discovered the fundamental law of physics upon which fluid power systems are based. Modern fluid power is made possible by pumps or compressors that create high pressures within a confined system, and then use relatively small quantities of that pressurized fluid (Figure 1.4).

(a)

(b)

(c)

(d)

FIGURE 1.1
The many uses of hydraulic applications: (a) Farming. (Courtesy Deere & Company, Moline, Illinois) (b) Aviation. (Courtesy Lockheed-Georgia Company, Marietta, Georgia) (c) Robot welding. (Courtesy Chrysler Corporation) (d) Road construction. (Courtesy Mannesmann Rexroth)

FIGURE 1.2
Example of large forces exerted by a front loader. (Courtesy Deere & Company, Moline, Illinois)

FIGURE 1.3
Early applications of fluid power.

FIGURE 1.4
An industrial hydraulic power unit provides a compact and efficient power system. (Courtesy Double A Products Company, Manchester, Michigan)

In 1795, Joseph Bramah, an Englishman, built the first hydraulic press (Figure 1.5). Yet, it was not until 1850 that the demands of the Industrial Revolution in Great Britain led to the fuller development of Bramah's water press and other industrial machines. At that time steam-driven water pumps powered the industrial presses. The development of heat engines, mechanical controls, and electricity during the latter part of the nineteenth century diverted attention from further developments and refinements of fluid power devices and systems.

Our modern fluid power era began recently, at the beginning of this century. In 1906, a hydraulic system replaced the electrical system for elevating and controlling guns in the battleship U.S.S. *Virginia*. In this initial installation, a variable speed hydrostatic transmission was installed to maneuver the guns. Today's ships make extensive use of hydraulic winches, controllable

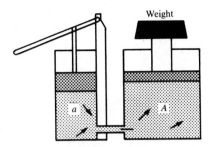

FIGURE 1.5
The 1795 Bramah hydraulic press had two different-sized pistons. Force applied to piston a was transmitted to piston A through pressurized fluid.

pitch propellers, elevators, missile launch systems, and navigation controls. The use of oil to replace water as a medium for transmitting energy played a critical role in the expansion of fluid power. Fluid power technology has advanced considerably since 1926, when the fluid power industry developed the self-contained, packaged system for industrial equipment, including a reservoir, pump, controls, and actuators (Figure 1.6). In little more than one-half century, fluid power technology has given rise to an important industry. In 1986, fluid power components and systems sales accounted for six billion dollars of the gross national product. With the increasing emphasis on automation, quality control, safety, and more efficient energy systems, this technology should continue to expand both within the United States and many other industrialized or emerging nations.

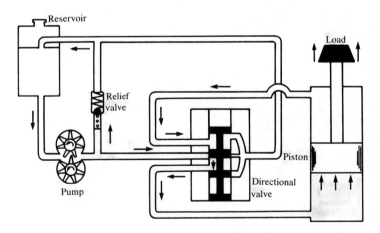

FIGURE 1.6
Components of a hydraulic system.

1.3 APPLICATIONS OF FLUID POWER

The systems design engineer has the option of choosing mechanical, electrical, electronic, pneumatic, or hydraulic methods of power transmission and control. Most choices require a combination of methods depending on the product and environmental requirements. These choices are guided by such criteria as efficiency, cost, safety, durability, and reliability. The widespread application of fluid power in most of our productive sectors, including aerospace, agriculture, communications, construction, defense, energy extraction, forestry, automated manufacturing, and transportation, illustrates the diversity and adaptability of fluid power systems to perform transmission and control functions. This extensive use indicates the wide range of career opportunities that will be available to those with knowledge, skill, and experience in fluid power technology.

Current developments in the growing field of robotics are adding more opportunity for engineers to design robots with high reliability and capacity to handle or process more demanding manufacturing process tasks. The most recent and sophisticated development is the use of programmable controllers in conjunction with hydraulic systems. A programmable controller is a digitally operated electronic apparatus that has a programmable memory for internal storage of instructions to implement such specific functions as logic, sequenc-

FIGURE 1.7
Space vehicle transporter. (Courtesy The Rexroth Corporation)

FIGURE 1.8
Agricultural hydraulic applications. (Courtesy Deere & Company, Moline,
Illinois)

ing, timing, counting, and arithmetic to control (through digital or analog
input/output modules) various types of machines or processes. Increasingly,
fluid power applications are being extended and influence our quality of life,
from reducing the need for manual work to the forms of recreation that become
available to us.

1.4 ADVANTAGES AND COMPARISONS OF FLUID POWER SYSTEMS

A designer familiar with fluid power recognizes the advantages of hydraulic
systems over mechanical or electrical systems for specific applications. Some of
these advantages are

- Multiplication of small forces to achieve much greater forces for performing work. Increasing the pressure or the area of the actuator allows the designer to increase the force exerted to a particular device.
- Accuracy of control of small or large forces with instant reversal is possible with hydraulic control systems.
- Constant force is possible in a fluid power system regardless of special motion requirements, whether the work output moves a few inches or several feet per minute. That is because rotational motion constant torque is possible over a few revolutions (as in the case of a hydraulic motor) or thousands of revolutions per minute.
- Simplicity, which means there are fewer moving parts than in comparable mechanical or electrical machines. Some systems are advanced in terms of the number and complexity of components and subsystems within a hydraulic circuit.
- Forces may be applied at a reasonable distance, at inaccessible points, or in any position—wherever conductors and actuators can be installed.
- Hydraulic parts are self-lubricating with the hydraulic liquid.

FIGURE 1.9
Hydraulic positioner. (Courtesy The Watt Car and Wheel Company, Barnesville, Ohio)

FIGURE 1.10
Hydraulically actuated robot for material handling. (Courtesy Amatrol)

- A complex sequence of operations can be controlled by a variety of mechanical devices such as cams, or electrical devices such as solenoids, limit switches, or programmable electronic controls.

Equipment and machine designers may select hydraulics over other systems for transmitting and controlling power because of one or more of the previous advantages. There are three major features of hydraulic equipment common to most applications

1. The efficient multiplication of forces through a liquid medium.
2. The flexibility of power transmission through a liquid medium.
3. The accurate control of motion of the machine's component parts.

Hydraulics provides almost unlimited applications in most of our major industries from agriculture to aerospace. The compatibility of hydraulics to machine requirements, and the wide range of components and systems available for adapting to particular applications have contributed to the rapid growth of fluid power technology. Yet, the actuators that convert fluid energy to mechanical energy are essentially a variation of linear or rotary motion.

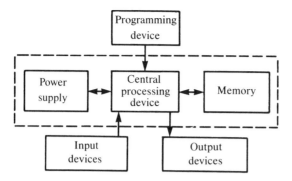

FIGURE 1.11
Block diagram of a microprocessor-based programmable controller.

FIGURE 1.12
Submersible rescue vessel. (Courtesy International Submarine Engineering, Limited)

FIGURE 1.13
Hydraulic retractable landing wheels. (Courtesy Lockheed-Georgia Company, Marietta, Georgia)

1.5 FUTURE PROSPECTS

While hydraulic components, seals, fluids, and conductors have undergone considerable design improvements, the most significant advances in hydraulic systems are being developed in the area of controls.

As electro-mechanical controls diversified the use of hydraulics, including servo-valves that use the feedback principle, remote sensing was limited previously as a source of information gathering.

Decision making and action response were rapidly expanded with the coupling of programmable control logic and production hydraulic control systems. Today's programmable controller makes automatic control systems much more efficient than the electromechanical systems of the past.

Electromechanical relays, counters, timers, and analog devices have been replaced with more reliable solid-state circuitry. The input/output modules provide the critical link between a programmable controller and the hydraulic devices.

Essentially the input-output modules electrically protect the programmable controller (PC) from dangerously high voltages. The low DC signal from the PC is applied to an electronic gate, which closes a switch to allow AC current to flow in the ouput devices circuits. The programmable controller then uses input signals, scans a user program in the logic, solves the ladder rungs, and keeps an operation running efficiently and safely.

Anticipated Programs

The combination of the three technologies—hydraulics, electricity, and electronics—has enabled designers of hydraulic systems to create more efficient, safe, and adaptive circuits. At the same time, while the systems are becoming more complex, integrated, and interdependent, increased instrumentation is necessary to locate, test, and correct malfunctioning components.

The increasingly refined controls applied to hydraulic systems will further augment the expansion of their use. During this century we have taken pride in creating high pressures to take advantage of the large forces we could apply. It is likely that the focus for the future will be directed at controlling energy transmission.

Summary

Hydraulics is used extensively in modern industry to perform work by transferring power via potential energy changes in a liquid flowing at a controlled rate. Hydraulic systems rely on a prime mover to drive the pump, such as an engine or an electric motor. Hydraulics applications have been rapidly expanded to many uses, ranging from very simple circuits to very complex and sophisticated systems found in aerospace and robots. Even though many of the scientific laws, principles, and concepts of fluid mechanics have been developed in the last three hundred years, the development of hydraulics is slightly more than a century old. Fluid power is an annual multibillion-dollar industry in the United States. Many other industrialized nations are actively developing and applying hydraulics in many economic sectors. Today, occupational careers ranging from fluid power mechanics, fluid power systems maintenance personnel, fluid power technicians, and fluid power engineers are available to persons with appropriate fluid power education and preparation.

There are numerous advantages to the use of hydraulics, such as the multiplication of forces, constant force or torque, simplicity, flexibility, and accuracy to control with reasonable safety. Fluid power applications can be found in such diverse sectors as aerospace, aviation, agriculture, construction, forestry, mining, transportation, communications, manufacturing, material handling, and recreation vehicles. As more people train to enter this technology and industry, hydraulics likely will become more complex and find even more diverse applications. The promise of expanded use of interfacing hydraulic components with programmable remote-sensing controls holds a very bright, exciting future for this important, growing technology.

Questions and Problems

1. Define the term *fluid power*.
2. What is the minimum list of components that would be required to construct a simple operating hydraulic circuit?
3. Why is a prime mover needed to create a functioning hydraulic circuit?
4. Examine the power steering on your car. Sketch the block circuit and label all major components.
5. List five advantages of hydraulics.
6. Prepare a one-page essay on the historical development of the hydraulics technology.

Suggested Learning Activities

Early in the course students should preview the videotape *Basics of Hydraulics* available from the National Fluid Power Association. An industrial plant tour that illustrates diverse applications of hydraulics is valuable in creating interest and motivation for studying this technology. This tour creates awareness of the hydraulics technology and career opportunities. Another suggested activity is to create three groups in the class to compare and contrast the advantages of mechanical systems, electrical systems, and hydraulic systems for transmitting and controlling power.

Suggested Readings

Decker, Robert L. *Hydraulics*. Mid-America Vocational Curriculum Consortium, Inc., State Department of Vocational and Technical Education, Stillwater, Okla. 1980.

Esposito, Anthony. *Fluid Power with Applications*. Englewood Cliffs, N.J.: Prentice-Hall, 1980.

Sullivan, James A. *Fluid Power: Theory and Applications,* 2nd ed. Reston, Va.: Reston, 1982.

Wolansky, William; Nagohosian, John; and Henke, Russell W. *Fundamentals of Fluid Power*. Boston: Houghton Mifflin, 1977.

CHAPTER 2

Fundamentals of Mechanics and Fluid Statics

Introduction

This chapter reviews basic laws of mechanics and gives the principal properties of a fluid at rest. How these fluid properties relate to hydraulic systems is explained. After studying this chapter, you will know how to calculate force, work, and power; understand fundamental fluid properties; and be able to work the problems at the end of the chapter.

The property of a fluid we usually first think of is its pressure, which is the force exerted across a unit area of the fluid. We therefore need to consider the laws of basic mechanics to define a force. A force is also the mechanism by which work is performed, so the power or rate of doing work can also be studied. In this chapter fundamental properties of fluids, such as viscosity, density, and Pascal's law, are examined.

Key Terms

ABS, =gAgE+14.7 *(handwritten annotation)*

Absolute Pressure: Pressure measured from absolute zero pressure (that is, from no pressure at all) rather than from atmospheric pressure (approximately 14.7 psi). Absolute pressure (psia) is equal to atmospheric pressure (patm) plus gauge pressure (psig). Thus, psia = 14.7 + psig for calculation purposes. A complete list of abbreviations and symbols in this chapter appears in Appendix B.

Absolute Temperature: Temperature measured from the absolute zero of temperature. Thus, °Rankine = °Fahrenheit + 460.

Acceleration: Rate of change of velocity.

Kinematic Viscosity: Viscosity divided by the mass density of the fluid. Most often measured in centistokes (cS), which have dimensions of centimeters (cm) and seconds (s).

Force: Will change the state or the velocity of the body upon which it is acting. Usually given in units of the pound (lb).

Mass: The amount of matter in a body. The standard unit of mass is the kilogram (kg). Another unit used is the pound mass, which one pound force will accelerate at the rate of 32.1739 ft/s^2 at sea level and 45° N latitude.

Mass Density: The amount of mass, m, in a given volume of material. Stated another way, $\rho = m$/volume.

Velocity: The rate of change of distance in a particular direction. The velocity of fluid flow is usually measured in feet per second (ft/s).

Viscosity: (sometimes called Dynamic Viscosity) A measure of the internal friction or the resistance of a fluid to flow, usually expressed in centipoise (cP).

2.1 FORCE

Newton's three laws of motion help to explain the concept of force. His second law, relating force and rate of change of momentum, is

$$\text{force} = (\text{mass})(\text{acceleration})$$

$$F = ma$$

where   force, F, is in lb
mass, m, is in slugs
acceleration, a, is in ft/s^2

Newton's third law, which states that action and reaction are mutually opposed, gives rise to the concept of load, or the overcoming of a resistive force. In addition to those due to acceleration, resistive forces can be frictional or inertial.

2.2 WORK, ENERGY, AND POWER

To perform work, a force must be overcome and moved through a distance. For this we write

1 PROBLEM

$$\text{work } (W) = \text{force (lb)} \times \text{distance (ft)}$$

We know work and energy are equivalent and that whatever happens, energy can be neither created nor destroyed, but may be converted from one form into another. For fluid power applications, the kind of energy that can best be utilized is pressure energy, although there are other forms: chemical, kinetic, potential, light, electrical, heat, and sound energies. Figure 2.1 shows examples of different forms of energy.

From Figure 2.1 we see that energy can also be defined at the ability to perform work. Figure 2.2 illustrates how the different forms of energy are interrelated and how energy is generated, distributed, and utilized.

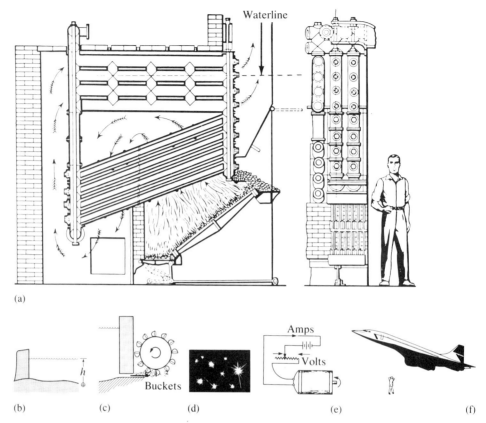

(a)

(b) (c) (d) (e) (f)

FIGURE 2.1
Some forms of energy: (a) First Babcock & Wilcox boiler, patented 1867.
(Courtesy Babcock & Wilcox) (b) Water within a reservoir. (c) Paddle wheel. (d)
Light from the stars. (e) Electric circuit. (f) Hearing the sound from a jet
engine.

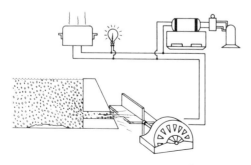

FIGURE 2.2
Related forms of energy.

2.3 POWER

To define power we must think of the rate of doing work. The usual way to express this quantity is to measure the rate in ft·lb/s. (Remember, 1 ft·lb is one unit of work.) Note also that 550 ft·lb/s is defined as one horsepower, or 1 hp.

2.4 TORQUE

Thus far we have considered linear motion. Fluid power applications are often concerned with rotary motion. Torque about a point is defined as the product of a force and the nearest distance of the line of action of that force to the point. An alternative way to define work done is to use work as the product of torque and angle turned through. Thus

$$\text{work} = \text{torque (lb·ft)} \times \text{angle turned through (radians)}$$

We can also calculate the power of rotation since it is represented by the product of torque and angular rate of rotation of the torque.

2.5 PROPERTIES OF A FLUID

A fluid is defined as a substance incapable of sustaining a shearing stress. This means that the fluid will assume the shape of the container it is in. Also, a fluid may be liquid or gaseous. Since we will not deal with transmission of power by means of gaseous pressure (pneumatics), we shall deal largely with liquid sub-

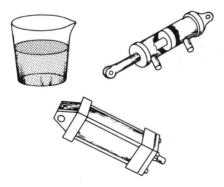

FIGURE 2.3
Liquids conform to their container shape.

stances and more usually with hydraulic oil. Occasionally we will work through examples with water or air to improve our basic knowledge of fluids. Note that only in extremely rare cases can a fluid sustain a tensile stress. Thus, a fluid usually sustains pressure in a compression mode only, but it can transmit force and power when totally enclosed. Because a liquid conforms to the shape of its container, we can pump liquids through pipes, tubing, hoses, and ports.

2.6 PASCAL'S LAW

Probably the most fundamental law in fluid power is that discovered by the French scientist Blaise Pascal. It is a two-part statement: *Pressure at any one point in a static fluid is the same in every direction, and pressure exerted in a confined fluid is transmitted equally in all directions acting with equal force on equal area.* To understand this fundamental law, consider two examples. First, imagine a lake 50 ft deep with a standpipe as shown in Figure 2.4. The pressure on the lake bottom is equal to $50 \times 0.433 = 21.67$ psi.

To generate the height of water required in the standpipe, the pressure at the base of the standpipe will have to be $h \times 0.433$ psi, which is exactly the pressure produced there as the result of hydrostatic action. Also note the pressure has been communicated in a horizontal manner along the connecting pipe, thus validating Pascal's law.

The second example considers the case of two cylinders connected by a pipe and filled with hydraulic oil (Figure 2.5). A 200 lb load is placed upon piston A in the left-hand cylinder, assumed to have a cross-sectional area of 2 in.^{2} As a result there will be an undiminished pressure throughout the oil equal to 100 psi. Looking at it another way, to support the load of 200 lb, a pressure of 100 psi has to be exerted by the oil on the piston surface. This pressure occurs as a direct result of Pascal's law.

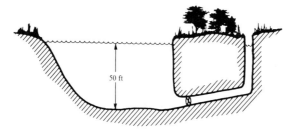

50 ft

FIGURE 2.4
Reservoir and standpipe.

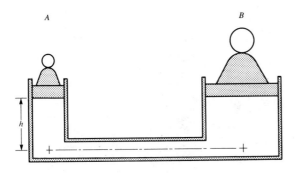

FIGURE 2.5
Pipe-cylinder combination.

Note that piston *B*, which has a cross-sectional area of 8 in.^{2}, will be able to support a load of 800 lb and be in equilibrium. Thus, a four-fold increase of force has been achieved by this simple device. Also observe that since energy can be neither created nor destroyed, piston *B* can move only one-fourth the distance moved by piston *A*. In this case, assuming the load on piston *B* is required to be moved by 2 in., we must have

work done *on* left-hand piston = work done *by* right-hand piston

or

force a × distance A = force b × distance B

$$200 \times \text{distance } A = 800 \times 2 \text{ in.} \cdot \text{lb}$$

$$\text{distance } A = 8 \text{ in.}$$

2.7 ABSOLUTE AND GAUGE PRESSURE

Until now we have been discussing pressures that exist in a fluid in addition to the all surrounding, pervading pressure of the earth's atmosphere. Usually this so-called gauge pressure is measured by some kind of pressure-measuring instrument in a pressurized fluid. One device used to measure pressure is the conventional Bourdon pressure gauge shown in Figure 2.6. Absolute pressure must be used to calculate changes of state of a gas, so an additional 14.7 psi have to be added to the gauge pressure.

With negative pressure, the same algebraic addition applies, which leads to an absolute pressure less than the pervading or atmospheric pressure. Thus, if we need 6.0 psi vacuum in a liquid, the absolute pressure in that liquid is equal to 14.7 − 6.0 = +8.7 psi absolute or −6.0 psi gauge.

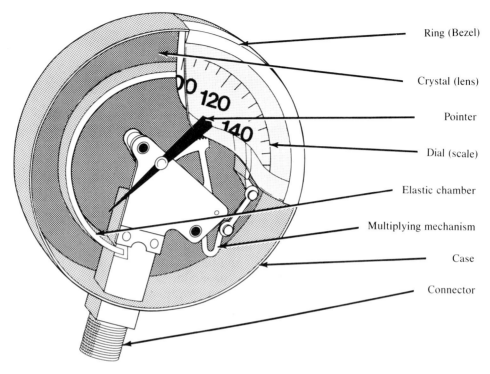

FIGURE 2.6
A Bourdon gauge. (Courtesy Marshalltown Instruments)

2.8 DENSITY AND PRESSURE

It will be shown later that an important property of a fluid is its density, or its mass per unit volume. It is more common to quote the so-called *specific weight,* which gives the weight per unit volume, designated by the symbol γ. We must know the weight and the volume of a quantity of fluid before we can determine its specific weight. To give some representative values, the specific weight for water is approximately 62.4 lb/ft^3, and the specific weight for air is approximately 0.075 lb/ft^3. As an example: One barrel of oil weighs 435 lb and contains 55 U.S. gallons. If the empty barrel weighs 25 lb, compute the specific weight of the oil, given 231 in.^{3} = 1 U.S. gallon.

$$\text{weight of oil } = 435 - 25 = 410 \text{ lb}$$

$$\text{volume of oil} = 55 \text{ gal} \times \frac{231 \text{ in.}^3}{1 \text{ gal}} \times \frac{1 \text{ ft}^3}{1728 \text{ in.}^3}$$

$$= 7.35 \text{ ft}^3$$

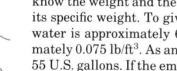

Therefore

$$\text{specific weight} = \frac{\text{weight}}{\text{volume}}$$

$$= \frac{410}{7.35} = 55.8 \text{ lb/ft}^3$$

The concept of specific weight will enable us to determine the pressure at the base of a column of fluid. For instance, imagine 1 ft³ of water (which weighs 62.4 lb) acting over an area of 1 ft². The pressure due to that 1 ft³ on the base of the cube, since it is 1 ft high, is 62.4 lb/ft², or 62.4/144 = 0.433 psi. In a similar way it can be shown that for a column 2 ft high (regardless of its cross-sectional area), the pressure on the base is 2 × 0.433 = 0.866 psi, and for any height h, the pressure will equal 0.433h psi. The pressure on the base of a column of any other fluid would be given by

| PROBLEM

$$p = 0.433 \times \text{sg} \times h \quad \text{(FT)}$$

where sg is the fluid specific gravity, which is the weight of the fluid divided by the weight of an equal volume of water. This analysis is generalized by stating that the pressure at the bottom of a container is equal to the product of the depth below the surface and the specific weight of the liquid.

The concept of pressure head is often useful and follows from the previous paragraph. *Pressure head* is defined as the depth of a liquid that causes a given pressure, and it is equal to the increase in liquid level in a standpipe due to pressure within the vessel to which it is attached.

Example 2.1 ▶ What is the depth of a freshwater lake if the pressure on the bottom is 8 psi?

Solution:

MIDTERM PROBLEM

$$p = 0.433(\text{sg})h \quad \text{(PSI)} \quad \text{(NO UNITS)}$$

$$h = \frac{8}{0.433 \times 1.0} = 18.5 \text{ ft}$$

◀

2.9 MEASUREMENT OF SMALL GAUGE PRESSURES

Pressure can be measured by using a Bourdon pressure gauge as shown in Figure 2.6. Other instruments include the piezometer and the mercury (or other liquid) manometer. The piezometer, which can be used only with liquids, is simply a small-bore, transparent wall tube with one end open to the atmosphere and the other connected to the point where pressure measurement is required. The fluid rises to a height equal to the pressure head, and therefore the piezometer is extremely sensitive, with small pressures giving relatively large deflections of the liquid level. It is inconvenient to use for pressures larger than approximately 8 lb/in.²

2.10 DYNAMIC VISCOSITY

The property that describes *thickness* or *resistance to motion* is called the viscosity of a fluid. It is probably the most important property when considering fluid power systems. The two types of flow possible, laminar and turbulent, are closely related to the fluid viscosity. To best understand this, consider laminar flow occurring between two flat plates. Under these conditions, flow takes place in laminae, or layers, and is at rest relative to the surface of each plate.

Figure 2.7 shows two flat plates separated by a distance h. The bottom plate is at rest, and the top plate of area A moves at a velocity v.

In this simple example, the velocity at any layer is proportional to the distance of the layer from the fixed plate. The force required to push a given plate at the speed v increases as the distance h between the plates decreases and as the viscosity of the oil increases. These observations lead to a more exact definition of viscosity in which the shear stress (force per unit plate area) is related to the velocity gradient. The symbol for viscosity is usually the Greek letter μ

$$\mu = \frac{F/A}{v/h}$$

A more general formula for defining viscosity is

$$\text{shear stress} = \mu \frac{du}{dy}$$

where u is the velocity at any point y

du/dy is the rate of shearing strain.

A mathematical analysis shows that the dimensions for viscosity are mass divided by length and by time. In the centimeter gram system (cgs) the unit is the *poise*. Thus, the shear stress is in dyne/cm². The velocity gradient is in $(\text{cm/s})(1/\text{cm}) = 1/\text{s}$. The viscosity is in gm/cm·s. A more common unit is the so-called *centipoise* (cP), or one-hundredth poise. By an extraordinary accident in historical definition of units, 1 cP turns out to be 1 millipascalsecond. Thus, the unit of force has been now included in the dimensions of viscosity.

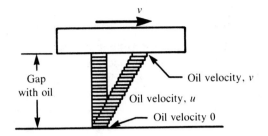

FIGURE 2.7
Velocity profile in an oil layer.

The conversion units are

$$\frac{1\ \text{lb}\cdot\text{s}}{\text{ft}^2} = \frac{1\ \text{slug}}{\text{ft}\cdot\text{s}} = 4880\ \text{cP}$$

In addition,

$$1\ \text{reyn} = \frac{1\ \text{lb}\cdot\text{s}}{\text{in.}^2}$$

therefore

$$1\ \text{poise} = 1.45(10^{-5})\ \text{reyns}$$

This last number is easily memorized, since $1 + 4 = 5$!

Example 2.2 ▶ Two parallel glass plates measuring 12 in. square are separated by oil having a specific gravity of 0.85 and kinematic viscosity equal to 60 cS. Calculate the force necessary to maintain a speed on the top plate of 30 mm/s, if the plates are separated by 0.2 in.

Solution: Assume linear velocity distribution. First, find the shear stress developed and multiply that shear stress by the area over which it acts.

$$\text{velocity gradient}\ = \frac{30\ \text{mm}}{\text{s}} \times \frac{1\ \text{cm}}{10\ \text{mm}} \times \frac{1}{0.2\ \text{in.}} \times \frac{1\ \text{in.}}{2.54\ \text{cm}} = \frac{5.906}{\text{s}}$$

$$\text{dynamic viscosity} = 0.85 \times \frac{1\ \text{gm}}{\text{cm}^3} \times \frac{60}{100} \times \text{cm}^2 = \frac{0.51\ \text{gm}}{\text{cm}\cdot\text{s}}$$

$$\text{shear stress}\quad = \mu \times \frac{v}{h} = 0.51 \times 5.906 = \frac{3.012\ \text{dynes}}{\text{cm}^2}$$

$$\text{force}\quad = \frac{3.012\ \text{dynes}}{\text{cm}^2} \times (12 \times 2.54\ \text{cm})^2 = 2800\ \text{dynes}$$

dyne   = the fundamental unit of force: A force applied to a mass of 1 gram for 1 s would give it a velocity of 1 cm/s. ◄

2.11 KINEMATIC VISCOSITY

In most calculations the quantity μ does not appear unless it is divided by the mass density. Thus, the quotient of dynamic viscosity and density is usually represented by the Greek letter nu (ν) and has dimensions of length squared divided by time. When the cgs system is used, the unit is the *stoke*, which has dimensions cm^2/s. Again, a one-hundredth part of the unit is used so that a convenient size may be obtained. Oil companies and oil manufacturers, therefore, quote results in terms of centistokes (cS).

Before we leave kinematic viscosity, you should recognize that since the specific gravity of oil is approximately 0.85 and the density of water in cgs units is unity (by definition), then 1 cP is approximately equivalent ot 1 cS for oil. In no other system is such a convenience enjoyed.

Example 2.3 ▶ Complete the following table which refers to a hydraulic oil. (Given: Dynamic viscosity = 0.1 gm/cm · s, specific gravity = 0.85)

Solution:

Dynamic Viscosity			Kinematic Viscosity		
gm/cm · s (poise)	kg/m · s	slug/ft · s	cm^2/s (stoke)	ft^2/s	m^2/s
0.1 (only value given)	0.01	$2.09(10^{-4})$	0.118	$1.269(10^{-4})$	$0.118(10^{-4})$

Deduction of numbers in other columns shown as follows:

$$\frac{0.1 \text{ gm}}{\text{cm} \cdot \text{s}} = 0.1 \left(\frac{\text{gm}}{\text{cm} \cdot \text{s}}\right)\left(\frac{1 \text{ kg}}{1000 \text{ gm}}\right)\left(\frac{100 \text{ cm}}{1 \text{ m}}\right) = 0.01 \left(\frac{\text{kg}}{\text{m} \cdot \text{s}}\right)$$

$$= 0.01 \left(\frac{\text{kg}}{\text{m} \cdot \text{s}}\right)\left(\frac{\text{slug}}{14.594 \text{ kg}}\right)\left(\frac{0.3054 \text{ m}}{1 \text{ ft}}\right) = 2.09(10^{-4}) \left(\frac{\text{slug}}{\text{ft} \cdot \text{s}}\right)$$

therefore

$$\nu = \frac{\mu}{\rho} = \frac{0.1}{0.85} = 0.118 \left(\frac{cm^2}{s}\right)$$

$$= 0.118 \left(\frac{cm^2}{s}\right)\left(\frac{1 \text{ m}}{100 \text{ cm}}\right)^2 = 0.118 \, (10^{-4}) \left(\frac{m^2}{s}\right)$$

$$= 0.118(10^{-4}) \left(\frac{m^2}{s}\right)\left(\frac{1 \text{ ft}}{0.3045 \text{ m}}\right)^2 = 1.269(10^{-4}) \left(\frac{ft^2}{s}\right)$$

◀

2.12 MEASUREMENT OF VISCOSITY

The actual value of viscosity is important. For instance, if an oil chosen for a fluid circuit has a viscosity value too high, this results in high resistance to flow, high consumption of power, and a resulting generation of high temperatures. If, however, the viscosity value is too low, there would be increased leakage past glands and seals, resulting in a pressure loss. There would also be excessive wear due to breakdown of oil films between adjacent moving parts.

Viscosity can be measured with a Saybolt Universal viscosimeter, as diagrammed in Figure 2.8. The idea behind this device is that the viscosity value is related empirically to the time required for a fixed quantity of liquid (60 ml) to flow through a standard orifice at a given head. The temperature at which the measurement takes place is carefully recorded for reasons described later. The time of flow is referred to as the Saybolt Universal Seconds (SUS) viscosity. Since a thick liquid flows slowly, the SUS viscosity will be higher for a "thick" liquid than for a "thin" liquid. The relationships between viscosity in cS and SUS are

$$cS = 0.226t - \frac{195}{t} \qquad t \le 100 \text{ SUS}$$

and

$$cS = 0.220t - \frac{135}{t} \qquad t > 100 \text{ SUS}$$

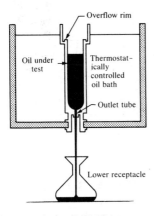

FIGURE 2.8
The Saybolt Universal viscosimeter.

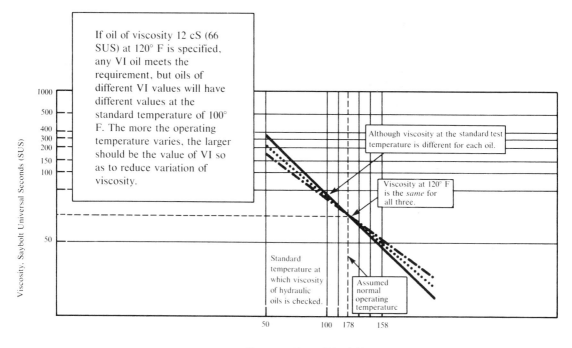

FIGURE 2.9
Variation of viscosity with temperature.

A second method of determining viscosity is to time the fluid flow through a capillary tube, again at a standard pressure head. The results for viscosity either in SUS or cS are usually obtained at 100° F and 210° F. When plotted in a logarithmic manner (*logarithmic* means each equal interval of the graph changes by a factor of 10) for the ordinate (*Y*-axis) against a linear scale of temperature on the abscissa (*X*-axis), the two points may be joined by a straight line. That line then represents a close approximation to the value of kinematic viscosity at any other temperature (see Figure 2.9).

2.13 PRESENTATION OF KINEMATIC VISCOSITY RESULTS (EFFECT OF TEMPERATURE)

By choosing a suitable vertical axis you can arrange to have viscosity values that vary in a straight-line manner when plotted against temperature. Knowledge of this variation in temperature is very important to the designer of agricultural and other machinery, and to operators of this machinery. Changes in temperature and in viscosity arise from the seasonal and geographical variation of atmospheric temperature, and also from the changes imposed on it by the machinery used with it. Thus, the temperature of oil used in a refrigeration compressor will be much cooler than that used in an internal combustion engine. The viscosity of the oil will therefore be significantly greater. In addition, the shearing action of the oil within these two temperature environments will cause energy to be delivered to the oil and increase its temperature. Again, this results in a reduction of viscosity or an effective "thinning down." Thus, the smaller the variation of viscosity with temperature, the easier it will be for the designer or operator to choose the correct oil for the particular application.

The change of viscosity with temperature requires a classification of the magnitude of this viscosity variation with temperature. In 1929 it was claimed that Gulf Coast oils would have the greatest variation, defined by an index 0. Pennsylvania oils were considered to have the least variation with temperature, and were designated as an index 100. All other oils were then defined by an index between these two extremes, as shown in Figure 2.9. The particular reference oils have to be chosen so that they and the oil under study have the same viscosity at 100° F.

Since the introduction of the index in 1929, many new additives and other oils have been produced so that better viscosity temperature characteristics are now available and give a VI (viscosity index) of greater than 100. A typical oil used in an agricultural tractor engine, for instance, has a VI of approximately 140. There are also artificial ways of improving the VI, such as adding high polymers. The effect on the viscosity of a small amount of additive is disproportionate to the change in VI produced, which can easily be as high as 75 points.

All that is generally needed to specify suitable oil for a particular machine is to quote a required viscosity at a given temperature. Suppose a kinematic viscosity of 12 cS is needed at a temperature of 120° F. Since 12 cS are equivalent to 66 SUS, we can draw the three lines representing the VI values of 0, 100, and 150 in Figure 2.9. Note that each line goes through our operational point, so we may select an oil with any value of VI for our particular task. Provided that the oil temperature remains at 120° F, the specifications will have been met. Let us now deduce the viscosity values we must have at the standard temperature of 100° F. We note that for an oil with a VI of 150 it will have to have a viscosity of 85 SUS. For a VI of 100 it will have to be approximately 88 SUS and 91 SUS for a VI of zero. Although the oil-operating temperature is seldom constant, the operating range is small (usually less than 50° F). It can be seen, however, that if a wide range of operating temperature is expected, the viscosity index value of the oil used assumes great importance.

Summary

This chapter contains the basics of applied mechanics. The fundamental concepts of force, pressure viscosity, work, energy, and power are given together with the manner in which these concepts are related to oil flow phenomena and fluid power technology. The evaluation and measurement of oil viscosity as explained in this chapter are important considerations.

Questions and Problems

1. Describe the three forms of energy involved in the generation, distribution, and utilization of electrical power. Show in logical sequence how each form is converted to the next.
2. A person weighing 160 lb walks upstairs from the second to the fifth floor. If each floor is 10 ft high, calculate the amount of work done in lb·ft.
3. In Problem 2, if it takes the person two minutes to walk up the stairs, calculate the average horsepower. (Given: 1 hp = 550 ft·lb/s)
4. A body weighs 50 lb/ft^3. If it is of volume 0.5 ft^3, calculate the acceleration generated in it, when a horizontal force of 5 lb is applied.
5. If in Problem 4 the force is removed after 5 s and the body continues to move at that velocity, compute the work done on the body (a) in 6 s at constant velocity; (b) in the first 15 s from rest.
6. How much torque is delivered by a 0.25 hp electric motor rotating at 1800 rev/min?
7. What is the potential energy of 10 ft^3 of water at a height of 50 ft?
8. The handle of a wrench is 14 in. long. (a) Calculate the torque exerted if you apply a 40 lb force on the end of the handle. (b) If you rotate the wrench through an angle of 30°, calculate the work done. (c) If you rotate the wrench through 360° in 5 s, calculate your rate of work in ft·lb/s.
9. A plate of area 2.5 ft^2 is placed horizontally in a water reservoir at a point 25 ft below its surface. Calculate the total force on one side of the plate.
10. In Figure 2.5 the left-hand limb has a diameter of 3 in. and the right-hand limb is 4 in. A weight of 450 lb is placed on piston A, and it is pushed down through a distance of 2 ft. What weight would the right-hand limb support? How much work has been performed in the movement?
11. In Problem 10, with reference to Figure 2.5, determine before the movement (a) the gauge pressure and absolute pressure at piston A; (b) the gauge pressure and absolute pressure at piston B; (c) the absolute pressure in the bottom limb if h = 30 ft and the specific gravity of the oil is 0.875.
12. Seventy-five gallons of a hydraulic fluid weigh 557 lb. (a) What is the mass of the oil? (b) What is its density? (c) What is the specific weight?
13. A can weighs 5 lb when empty, 55 lb when filled with water, and 68 lb when filled with glycerine. What is the specific gravity of glycerine?

14. Complete the following table, which refers to a relatively heavy lubricating oil of specific gravity 0.89.

Dynamic Viscosity				Kinematic Viscosity		
gm/cm · s (poise)	slug/ft · s	lb · s/in.^{2}	kg/m · s	cm^2/s (stokes)	ft^2/s	m^2/s
1.30						

15. Some of the oil in the table of Problem 14 is poured onto a horizontal surface, and a piece of metal with a base area of 0.25 m^2 is placed upon it. If a film of oil of thickness 0.5 mm is produced and the steel rides upon this film with a speed of 0.8 m/s, calculate the forces resisting motion. State your assumptions.
16. Two parallel plates each 3 ft. square are submerged 0.01 in. apart in water. If the viscosity of the water is 1 cP and a force of 1 lb is applied to one of the plates parallel to it, what relative velocity will be maintained?

Suggested Learning Activities

1. Specific weight of water and oil. Weigh a clean, open-top can. Fill it to the brim with oil. Weigh the can full of oil and pour the oil back into its original container. Clean the interior of the can with a cloth. Fill it with water and weigh again. From your results compute (a) the specific weight of water, (b) the density of water (choose two sets of units), (c) the specific gravity of the oil chosen.
2. Approximate variation of viscosity of water with temperature. Take the can in Activity 1 and pierce the bottom with a ⅛ in. diameter nail. Push a drinking straw through the hole, ensuring it is fixed in position with the top opening 1 in. above the bottom of the can. Fill with water at room temperature (measure the water temperature) and time how long it takes to empty. Fill with water at high temperature (place your finger over the hole until you have taken the temperature), and then again time how long it takes to empty. Plot a curve of the reciprocal of time to empty versus temperature. Compare the shape of the curve with the known pattern of variation. If you can, repeat this procedure with oil.

Suggested Readings

Sullivan, James A. *Fluid Power: Theory and Applications*, 2nd ed. Reston, Va.: Reston, 1982.

Wolansky, William; Nagohosian, John; and Henke, Russell W. *Fundamentals of Fluid Power*. Boston: Houghton Mifflin, 1977.

CHAPTER 3 ▬▬▬▬▬

Fluids in Motion

Introduction

This chapter explains how to calculate the properties of fluids in motion. It shows how the principles of conservation of mass flow and energy are used, and illustrates the effect of viscosity on driving fluids through pipes. Once this chapter has been understood, you will be able to calculate velocities and pressures in a moving fluid, calculate pressures and powers required to pump fluids at given rates, and solve the problems at the end of the chapter.

Bernoulli Equation: The sum of energy per unit weight of fluid flow due to pressure, velocity, and elevation.

Continuity: The property of constant mass flow rate or constant volumetric flow rate at any point along the duct or pipe.

Discharge Coefficient: The ratio of the actual flow rate to the theoretically calculated flow rate, assuming no energy loss.

Elevation Head: The quantity of energy of the fluid per unit weight due to fluid elevation, which has the dimensions of length.

Flow Area: The cross-sectional area of the pipe or duct through which the fluid flows.

Flow Rate: The volumetric displacement of fluid across any given point along the flow.

Laminar Flow: Steady flow taking place in fluid layers.

Pressure Head: The amount of energy of the fluid per unit weight due to pressure, which has dimensions of length.

Reynolds Number: Dimensionless ratio of the inertia to viscous forces in the fluid.

Specific Weight: The weight of unit volume of fluid.

Streamline: An imaginary line drawn everywhere tangential to the fluid direction.

Stream Tube: An imaginary envelope of streamlines through which fluid flows. Flow across the walls of such a tube is defined as not possible.

Turbulent Flow: Flow containing eddies and erratic, irregular components superimposed upon a mean velocity.

Velocity Head: The quantity of energy of the fluid per unit weight due to velocity, which has dimensions of length.

Venturi, Orifice: Devices to measure fluid flow rate by locally reducing the flow area, thus increasing the velocity and reducing pressure. (Velocity is a function of the square root of the measured pressure difference between two specific points.)

3.1 THE CONTINUITY EQUATION AND STEADY FLOW

The challenge of calculating motion properties is greatly reduced when considering steady flow. This means the pipe or conductor is filled with fluid, and as much liquid enters one end of the pipe as discharges from the other end within the same time. This phenomenon is called *continuity*. To help understand it, think in terms of a stream tube. Such a tube is a "bundle" of streamlines around the edge of an imaginary tube in the flow space. A streamline has the unique property that flow of fluid across it is not possible. Thus, fluid cannot cross the surface of a stream tube. The similarity between conventional flow (where oil usually does not leak through the pipe walls) and flow in a stream tube can be readily observed.

This useful concept is illustrated in Figure 3.1. It is also assumed the fluid within the stream tube occupies the same volume whatever the pressure is (i.e., it is incompressible). The quantity of flow, Q, or volumetric flow per unit time is therefore the same at any point along the pipe.

The actual value of Q also may be calculated by multiplying the average velocity in the pipe by the pipe cross-sectional area. Thus,

$$\text{volumetric flow rate} = Av$$

or

$$Q = A_1 v_1 = A_2 v_2 \tag{3.1}$$

This equation is known as the continuity equation. Usually, the flow is given in gal/min, ft³/min, and the area in in.² However, it is more scientific to use flow in ft³/s and area in ft², resulting in average velocity being in units of ft/s. We may

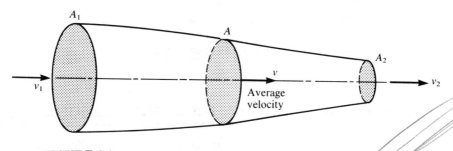

FIGURE 3.1
Stream tube with fluid flowing at a steady rate.

also use the SI system, in which Q is expressed in m^3/s and velocity is expressed in m/s.

Example 3.1 ▶ Oil flows along a horizontal pipe at the rate of 3.5 ft^3/min. Compute the average velocity of the oil where the diameter of the pipe is (a) 1.5 in. (b) 0.75 in. Also calculate the mass flow per second if the oil used is MIL-H-5606B.

Solution: The diameter of the bore (and also of the piston) is often given in inches, and the area as $A = \pi r^2$. But $r = D/2$ and, therefore, $A = \pi D^2/4 = 0.7854D^2$. At point 1,

$$A_1 = 0.7854(1.5)^2 \text{ (in.}^2) \text{ (1 ft}^2/144 \text{ in.}^2) = 0.01227 \text{ ft}^2$$

Similarly, $A_2 = 0.00307 \text{ ft}^2$. (Note that when $D_2 = D_1/2$, then $A_2 = A_1/4$).

$$Q = 3.5 \text{ (ft}^3/\text{min)} \text{ (min/60 s)} = 0.05833 \text{ ft}^3/s$$

Using the continuity equation, the average velocities can be computed as

$$v_1 = \frac{0.05833 \text{ ft}^3/s}{0.01227 \text{ ft}^2} = 4.75 \text{ ft/s}$$

$$v_2 = 4v_1 = 19.0 \text{ ft/s}$$

From Appendix A we find the density of MIL-H-5606B is 1.64 $slug/ft^3$, thus mass flow rate

$$M = \text{velocity} \times \text{cross-sectional area} \times \text{mass density}$$
$$= 4.75 \times 0.01227 \times 1.64$$
flow rate $M = 0.096$ slug/s

If conversion from gallons per minute (gpm) to pipe velocity in feet per second (ft/s) is required, the internal cross-sectional area of the pipe, A, has to be used. Thus, from Appendix C

$$v = 0.3208Q/A = 0.408Q/D_i^2 \tag{3.2}$$

where v = fluid velocity (ft/s)
Q = flow rate in gpm
D_i = internal pipe diameter (in.) ◀

3.2 CONSERVATION OF ENERGY (THE BERNOULLI EQUATION)

The concept of energy was discussed in Chapter 2. There it was stated that although one form of energy can be converted to another form, it cannot be

TABLE 3.1
Energy components in a flowing fluid.

	Point 1	Point 2
Potential Energy	h_1	h_2
Pressure Energy	$p_1/\rho g$ or p_1/γ	$p_2/\rho g$ or p_2/γ
Velocity Energy	$v_1^2/2g$	$v_2^2/2g$

created or destroyed. For fluid flowing in a pipe, this concept of constant flow energy results in a formula. This formula adds pressure energy, velocity energy, and height energy, and considers the sum of these three terms as a constant, provided there are no losses due to friction. Figure 3.2 illustrates the components of the Bernoulli equation.

Table 3.1 computes the quantities that represent the energy components. For the system with no energy losses, we add each column and equate the sum in each case. The three terms of each side of the equation are described as elevation head, pressure head, and velocity head; thus

$$h_1 + \frac{p_1}{\gamma} + \frac{v_1^2}{2g} = h_2 + \frac{p_2}{\gamma} + \frac{v_2^2}{2g} \qquad (3.3)$$

where h = height above a given datum plane (ft)
p = pressure (lb/ft^2)
v = velocity (ft/s)
g = acceleration due to gravity (ft/s^2)
γ = the specific weight of the fluid = mass density $\times g$ (lb/ft^3)

For British units

$$\gamma = 62.4 \times \text{sg} \ (\text{lb/ft}^3)$$

3.3 POWER REQUIRED TO DRIVE FLUIDS

The previous section shows how, in the absence of energy losses, to determine the appropriate conditions at one point in a pipe if we know the conditions at other points in the flow. However, no flow takes place without losses. Figure 3.2 illustrates that friction would normally be expected to occur at the walls of the pipe. In addition, it shows by diagram that it is possible to supply the flow with further energy by means of a pump and to remove some energy from the flow with a motor or turbine. We envision the energy input from the pump in this diagram as a head gain, and the energy absorbed by the motor as head loss. Motors and turbines are constructed so they can utilize the energy in a fluid

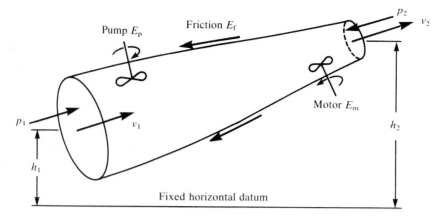

FIGURE 3.2
Energy components in fluid flow.

stream and apply power where required. We can, therefore, write an extended equation

$$h_1 + \frac{p_1}{\gamma} + \frac{v_1^2}{2g} + E_p = h_2 + \frac{p_2}{\gamma} + \frac{v_2^2}{2g} + E_m + E_f \qquad (3.4)$$

where the subscript p represents the energy gain due to a pump, and the subscripts m and f represent the energy loss due to a motor and due to friction.

Example 3.2 ▶ At one pipe location in the flow of water, the pressure is 25 psi and the speed is 1000 ft/min. At another point in the pipe, with a vertical location 16 ft above the first point, the pressure is 15 psi and the speed is 1400 ft/min. Calculate the head lost by friction in the pipe. What pressure loss in psi does this loss correspond to?

Solution: Write the total energy at each point using the extended Bernoulli equation, Equation 3.4.
At the first location

$$\frac{p_1}{\gamma} + v_1^2 g + h_1 = H_1$$

Assume the flow datum (horizontal plane) is at the level of the first location. We must use absolute units (ft and s). Thus

$$\begin{array}{l}\text{total energy} \\ \text{(or "head")}\end{array} = 25\left(\frac{\text{lb}}{\text{in.}^2}\right)\left(\frac{144\ \text{in.}^2}{1\ \text{ft}^2}\right) \times \underset{\underset{\text{pressure "head"}}{\uparrow}}{\frac{1}{62.4}\left(\frac{\text{ft}^3}{\text{lb}}\right)} + \left[1000\left(\frac{\text{ft}}{\text{min}}\right)\left(\frac{\text{min}}{60\ \text{s}}\right)\right]^2 \times \underset{\underset{\text{velocity "head"}}{\uparrow}}{\frac{1}{2(32.2)}\left(\frac{\text{s}^2}{\text{ft}}\right)} + \underset{\underset{\text{elevation "head"}}{\uparrow}}{0}$$

$$= (57.7 + 4.3 + 0)\ \text{ft}$$

40

$$H_1 = 62.0 \text{ ft}$$

Similarly, at the second location

$$\frac{\text{total energy}}{\text{(or "head")}} = \frac{15\ (144)}{62.4} + \left(\frac{1400}{60}\right)^2 \left(\frac{1}{2\ (32.2)}\right) + 16.0 + \underset{\underset{\text{friction "head" loss}}{\uparrow}}{E_\text{f}}$$

$$= (34.6 + 8.5 + 16.0) \text{ ft} + E_\text{f}$$

$$H_2 = 59.1 + E_\text{f}$$

Therefore, after equating the total head at each point ($H_1 = H_2$), the pressure head loss due to friction E_f is given by $62.0 = 59.1 + E_\text{f}$ or $E_\text{f} = 2.8$ ft of water.
From definition of pressure head, this loss is equal to

$$p_\text{f} = \gamma E_\text{f} = \frac{62.4\ (2.8)}{144} = 1.2 \text{ psi}$$

We now have to calculate the power required to pump the fluid through the pipe system. To do this we simply return to our basic equations of mechanics. Chapter 2 showed that

$$\text{work} = \text{force} \times \text{distance}$$

Also, the power is equal to the work done per unit time. In a pipe, the force is equal to the pressure times pipe area. Thus

$$\text{power} = \text{work } /\text{s} = (\text{pressure}) (\text{area} \times \text{distance/s})$$

So we can write

$$\text{power} = \text{pressure} \times \text{flow} \tag{3.5}$$

Now we calculate the amount of power required to propel the water through the pipe in the example, assuming the pipe has a diameter of 4 in. at the lower location.

Solution: The pressure to be generated includes the friction loss in the pipe plus the additional pressure due to the increase in elevation head (16 ft).

Thus

$$\text{pressure} = 62.4\ (2.8 + 16) = 1173.1 \text{ lb/ft}^2$$

$$\text{flow rate} = \frac{1000 \text{ ft}}{\text{min}}\left(\frac{1 \text{ min}}{60 \text{ s}}\right) \times \frac{\pi}{4}\left(\frac{4}{12}\right)^2 \text{ft}^2 = 1.45 \text{ ft}^3/\text{s}$$

$$\text{power} = \frac{1173.1 \text{ lb}}{\text{ft}^2} \times \frac{1.45 \text{ ft}^3}{\text{s}} = 1700 \text{ ft} \cdot \text{lb/s}$$

$$\frac{231 \text{ in}^3}{\text{gal}} \times \frac{\text{FT}^3}{1728 \text{ in}^3} \times \frac{\text{MIN}}{60 \text{ s}}$$

It is standard to quote flow rates in gallons per minute (gpm) and pressures in psi. Therefore, from James Watt's definition that one horsepower is equivalent to 550 ft · lb/s, the power required to pump the water in horsepower is

$$HP = \frac{pQ}{1.714} \qquad (3.6)$$

where p = the pressure drop in thousands of psi
          Q = the flow rate in gpm

It would be a good idea here to check this formula with the previous result.

$$\text{pressure} = \frac{1173.1 \text{ lb}}{\text{ft}^2} = 8.15 \text{ psi}$$

$$\text{flow rate} = \frac{1.45 \text{ ft}^3}{\text{s}} \times \frac{60 \text{ s}}{1 \text{ min}} \times \frac{7.48 \text{ gal}}{1 \text{ ft}^3} = 650.8 \text{ gpm}$$

$$\text{power} = \frac{(8.15)(650.8)}{(1000)(1.714)} = 3.094 \text{ hp}$$

Thus

$$\text{power} = 1700 \text{ ft} \cdot \text{lb/s}$$

As expected, the result is the same as before.

Conversely, we can determine from the previous analysis the pump head that has to be generated if a pump of known fluid horsepower output is given. Thus

$$H_{\text{p}} = \frac{3950 HP}{Q \text{ (sg)}} \qquad (3.7)$$

where (see Appendix C) HP = fluid horsepower generated by the pump
                              Q = pump flow, gpm
                              sg = specific gravity of fluid being pumped
                              H_{p} = pump head (ft). (Equivalent to E_{p} in Figure 3.2 and Equation 3.4.) ◀

3.4 APPLICATIONS OF THE BERNOULLI EQUATION

We have learned from Equation 3.3 that with no losses, the total flow energy in a duct remains constant. If we increase either the velocity head or the elevation head, conservation of energy principles will cause a reduction in pressure. One convenient way to measure flow rate is with a Venturi meter as shown in Figure 3.3.

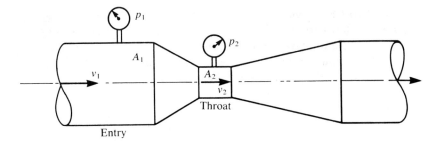

FIGURE 3.3
The Venturi meter.

In the Venturi, the diameter of the pipeline carrying the fluid is reduced in a smooth conical manner to the so-called *throat diameter*. A gradual enlargement is then achieved with a conical expansion to the original pipeline diameter. We can write the Bernoulli equation between points 1 and 2, assuming the velocities to be uniform across the pipe at these points, as

$$\frac{p_1}{\gamma} + \frac{v_1^2}{2g} = \frac{p_2}{\gamma} + \frac{v_2^2}{2g} \tag{3.3a}$$

(Note the elevation head is assumed to be the same at points 1 and 2. Therefore, they cancel from each side.)

We can also write the continuity condition as

$$A_1 v_1 = A_2 v_2 \tag{3.1}$$

From Equation 3.3a

$$\frac{v_2^2}{2g}\left[1 - \left(\frac{v_1}{v_2}\right)^2\right] = \frac{p_1 - p_2}{\gamma} \tag{3.8}$$

Upon substituting Equation 3.1 into Equation 3.8, we can write the velocity and flow rate as a function of pressure differences. Thus

$$v_2 = \sqrt{\frac{2(p_1 - p_2)}{\rho[1 - (A_2/A_1)^2]}} \tag{3.9}$$

and, since $Q = v_2 A_2$

$$Q = A_2\sqrt{\frac{2(p_1 - p_2)}{\rho[1 - (A_2/A_1)^2]}} \tag{3.10}$$

Since v_2 is greater than v_1, p_1 must be greater than p_2. This important principle is used in the carburetor of an automobile engine and is used extensively to measure flow rate of fluids. In the carburetor the sucking action applied to the vaporized fluid draws in atmospheric air for combustion.

Note that Equation 3.10 assumes no energy losses. In practice there is a small loss, so Equation 3.10 is modified to

$$Q = C_d A_2 \sqrt{\frac{2(p_1 - p_2)}{\rho[1 - (A_2/A_1)^2]}} \tag{3.11}$$

where C_d is known as the Venturi discharge coefficient and has a value between 0.96 and 0.99 for well-designed Venturis.

The pipes are usually circular in cross section, so

$$Q = C_d \left(\frac{\pi D_2^2}{4}\right) \sqrt{\frac{2(p_1 - p_2)}{\rho[1 - (D_2/D_1)^4]}} \tag{3.11a}$$

Note that the *fourth* power of diameter is involved here!

Another (less expensive) way of measuring flow rate (which incidentally leads to more installational convenience) is with an orifice meter, shown in Figure 3.4.

In this device a sharp edged orifice is placed in the flow at a convenient flange joint in the piping system. The difference in pressure between a point in distance D_1 upstream and a point in distance $D_1/2$ downstream of the orifice is measured. Although the analysis is the same as for the Venturi, the value C_d for the orifice is approximately 0.65, since there is clearly considerably more energy loss.

Example 3.3 ▶ You know the oil flow rate in a main pipe of 2 in. diameter is 300 gpm. At one point in the pipe the diameter is reduced to 1 in. Calculate the pressure difference between the 1 in. throat and the main flow, assuming no fluid energy loss and specific gravity of the oil of 0.86. Is the pressure at the throat larger or smaller than that in the main 2 in. pipe?

Solution: From Equation 3.2

$$v_2 = 0.408 \left(\frac{300}{1^2}\right) = 122.4 \text{ ft/s}$$

$$\text{fluid density} = \frac{(0.86)\,(62.4)}{32.2} = 1.67 \, \frac{\text{slug}}{\text{ft}^3}$$

From Equation 3.10, where $A_2 = \frac{\pi}{4}\left(\frac{1}{12}\right)^2 = 0.005454 \text{ ft}^2$

$$Q = v_2 A_2 = (122.4)\,(0.005454) = 0.005454 \sqrt{\frac{2(p_1 - p_2)}{1.67[1 - (\frac{1}{2})^4]}}$$

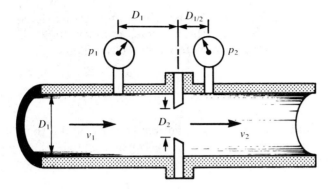

FIGURE 3.4
The orifice meter.

Square each side of the equation and transpose

$$\frac{14{,}981.8(1.67)0.9375}{2(144)} = p_1 - p_2$$

or

$$p_1 - p_2 = 81.4 \text{ psi}$$

Since a square root sign is involved in the calculation of $p_1 - p_2$, it is not obvious from the formula used which will be the higher pressure, p_1 or p_2. However, since the velocity in the 1 in. pipe is four times that in the main flow, the pressure there will be reduced due to conservation of energy. Thus p_1 will be greater than p_2. ◄

3.5 TYPES OF FLUID FLOW IN PIPES

In Chapter 2 the phenomenon of viscosity was explained. Because viscosity of a fluid is finite, resistance has to be overcome to generate flow in a pipe. You also learned in Chapter 2 that two types of flow are possible: laminar and turbulent. Both types of flow can and do exist in fluid power systems. There is no general analytical treatment of fluid flow properties. The forces actually affecting the flow are due to gravity, buoyancy, surface tension, cavitation, and electric and magnetic fields. Also, forces are a function of viscosity and inertia. Fortunately these are the two properties that are dominant under most flow conditions. To give an idea of the complexity of fluid flow (although it is viscosity that causes frictional resistance), the flow can only be analyzed by assuming it is either viscosity or inertia dominated. There is no analytical solution for turbulent (or inertia-dominated) flow. Under these conditions there exist only working formulas (or empirical results).

45

The quantity that describes the relative significance of forces due to inertia and due to viscosity is called the Reynolds number (Re). It is actually the dimensionless ratio of inertia force to viscous force. Thus

$$\text{Re} = \frac{\rho \bar{v} a}{\mu} \tag{3.12}$$

where ρ is the fluid density, $\bar{v}$ is the average velocity, and μ the viscosity. The quantity a is a characteristic length and is considered to be the pipe diameter, D. Thus in pipe flow, the type of flow is characterized by the quantity

$$\text{Re} = \frac{\rho \bar{v} D}{\mu} = \frac{4 \rho Q}{\pi \mu D} \tag{3.12a}$$

Another expression for Reynolds number using more convenient units is

$$\text{Re} = \frac{7740 \bar{v}(\text{ft/s}) \; D(\text{in.})(\text{sg})}{\mu(\text{cP})} \tag{3.12b}$$

and when kinematic viscosity is known

$$\text{Re} = \frac{7740 \bar{v}(\text{ft/s}) \; D(\text{in.})}{\nu(\text{cS})} \tag{3.12c}$$

When Re is less than 2000, laminar flow probably exists. In this type of flow, orderly, smooth parallel-line fluid motion is seen. Inertia-dominated (or turbulent) flow is characterized by irregular, erratic, eddy-like paths of the fluid particles and will usually occur if Re is greater than 4000. In the range of Re between 2000 and 4000, the flow is generally referred to as "in transition." Figure 3.5 diagrams the difference between laminar and turbulent flows, with both flows having the same average velocity.

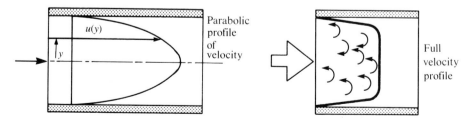

FIGURE 3.5
Laminar and turbulent flows in a pipe.

Example 3.4 ▶ A hydraulic oil of viscosity 105 cP and specific gravity 0.88 is flowing in a pipe of 1.25 in. diameter at a rate of 45 gpm. Would you expect the flow to be laminar or turbulent?

Solution: From Equation 3.2

$$v = 0.408Q/D^2 = 0.408(45)/(1.25)^2$$

$$v = 11.75 \text{ ft/s}$$

From Equation 3.12c

$$\text{Re} = 7740 \ (11.75) \ 1.25 \ (0.88) \ / \ 105 = 952$$

Since the Reynolds number is less than 2000, the flow should be laminar. ◀

3.6 FRICTION LOSSES

The energy loss in a pipe system consists of losses in the straight pipe portions and losses in fittings, bends, expansions, and contractions. Figure 3.6 shows these types of loss, where it is sufficiently accurate just to add all the friction effects of the individual pieces of straight pipe and individual fittings.

For the straight pipes we use the Darcy-Weisbach equation for head loss, H_L

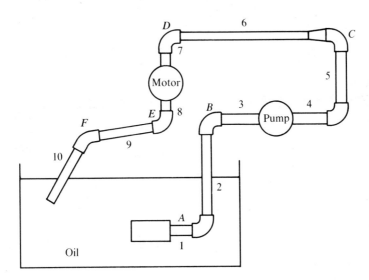

FIGURE 3.6
Typical hydraulic system energy loss components.

$$H_L \text{ (pipe)} = f\left(\frac{L}{D_i}\right)\frac{v^2}{2g}$$ (3.13)

where
f = friction factor
L, D_i = length and inside diameter of pipe
v = average fluid velocity
g = acceleration due to gravity

Therefore, in Figure 3.6, pipes 1, 2, 3, and 4 through 10 may be regarded as individual straight pipes and their combined effect added linearly. (Laminar and turbulent flows have different values of f. Methods of determining these quantities are given in the following sections.) In the case of pipe fittings, designated A, B, and C through H in Figure 3.6, the head loss may be expressed as

$$H_L \text{ (fittings)} = K_L v^2/2g$$ (3.14)

where values of K_L have been determined by experiment and are shown in Table 3.2. In addition, the approximate equivalent length of pipe of the fitting is tabulated in terms of numbers of pipe diameters. To determine this equivalent length a value of f must be assumed. A mean value of 0.02 has been assumed here with results sufficiently accurate for most purposes.

Laminar Pipe Flow Resistance. For laminar flow, usually considered to be where Re is less than 2000, the value of f for a relatively long pipe is given by

$$f = 64/\text{Re}$$ (3.15)

TABLE 3.2
Resistance factors for common valves and fitting.

Types of Fitting	K_L	Equivalent Pipe Length (diameters) (for $f = 0.02$)
ball check valve	4.0	200
gate valve, wide open	0.19	10
3/4 open	0.90	45
1/2 open	4.5	225
1/4 open	24.0	1200
globe valve, open	10.0	500
1/2 open	13.0	650
return bend	2.2	110
standard tee	1.8	90
90° elbow	0.8	40
40° elbow	0.4	20

If we now substitute Equation 3.15 into Equation 3.13, we arrive at the theoretical Hagen-Poiseuille equation, valid only for laminar flow.

$$H_L = \frac{64}{Re} \times \frac{L}{D} \times \frac{v^2}{2g} \tag{3.16}$$

Example 3.5 ▶ For the system of Example 3.4, calculate the head loss and the pumping power required for a 200 ft length of pipe.

Solution: From Equation 3.16

$$H_L = \frac{64}{952} \times \frac{(200)(12)}{1.25} \times \frac{11.75^2}{2(32.2)} = 277 \text{ ft of oil}$$

From transposed Equation 3.7

$$HP = \frac{Q(sg)H_L}{3950} = \frac{(45)(0.88)(277)}{3950}$$

$$P = 2.78 \text{ hp} \qquad \blacktriangleleft$$

Turbulent Pipe Flow Resistance. For turbulent flow, no theoretical analysis is possible. Values of f are based on experimental observations and are a function of the pipe wall roughness in addition to Re. However, most hydraulic circuits are fabricated from drawn steel or copper tubing. Therefore, their roughness is on the order of 5 millionths of an inch. For the purpose of these calculations, we assume the pipes are smooth-walled. Here, for Reynolds numbers between 4000 and 100,000 Blasius produced the result

$$f = 0.3164/(Re)^{0.25} \tag{3.17}$$

From Equations 3.12a, 3.13, and 3.17 we may obtain an expression for pressure loss (in lb/ft^2) per unit pipe length as

$$\frac{p_1 - p_2}{L} = 0.2414\left(\frac{\rho^{0.75}\mu^{0.25}\,Q^{1.75}}{D_i^{4.75}}\right) \tag{3.18}$$

where $p_1 - p_2 =$ pressure loss in length L (ft)
$\rho =$ density (slugs/ft^3)
$\mu =$ viscosity (lb · s/ft^2)
$Q =$ flow rate (ft^3/s)
$D_i =$ pipe inner diameter (ft)

For Reynolds numbers higher than 100,000, Prandtl gives a more complex relation

$$\frac{1}{\sqrt{f}} = 2 - \log_{10}(\text{Re}\sqrt{f}) - 0.8 \qquad \textbf{(3.19)}$$

Notice that for these higher Reynolds numbers it is not possible to develop an explicit relationship for pressure drop similar to Equation 3.18. Yet, with the range of flow rates and oils conventionally used, Reynolds numbers of this magnitude are seldom achieved. For a more complete analysis of how the friction factor varies with Reynolds number and roughness, refer to any of the Suggested Readings.

Example 3.6 ▶ Calculate the pressure required between the two ends of a pipe which is 2 in. in diameter and 100 ft in length, if the average oil velocity is 12 ft/s. Assume the oil being pumped has properties identical to that in Example 2.3.

Solution: First calculate Reynolds number. From Equation 3.12c, recalling that 1 cS = one hundredth of a stoke

$$\text{Re} = \frac{7740(12)2.0}{11.8} = 15,742$$

The flow is turbulent; thus, we use Equation 3.18, in which

$$\text{Re} = \frac{7740(12)2.0}{11.8} = 15,742$$

$$\rho = 0.85 \left(\frac{62.4}{32.2}\right) = 1.65 \,\frac{\text{slug}}{\text{ft}^3}$$

$$\mu = 2.09(10^{-4}) \,\frac{\text{slug}}{\text{ft} \cdot \text{s}} = 2.09(10^{-4}) \,\frac{\text{lb} \cdot \text{s}}{\text{ft}^2}$$

$$Q = 12\left(\frac{\pi}{4}\right)\left(\frac{2}{12}\right)^2 = 0.26 \,\frac{\text{ft}^3}{\text{s}}$$

$$D = \frac{2}{12} = 0.167 \text{ ft}$$

$$\frac{p_1 - p_2}{100} = 0.2414 \times \frac{(1.65)^{0.75}[2.09(10^{-4})]^{0.25}(0.26)^{1.75}}{(0.167)^{4.75}}$$

$$p_1 - p_2 = 1987.5 \,\frac{\text{lb}}{\text{ft}^2} = 13.8 \text{ psi} \qquad ◀$$

Example 3.7 ▶ Laminar flow is taking place through a pipe of 1 in. diameter at the rate of 6 gpm. This pipe is part of a pipe system with two wide open gate valves, four 90° elbows, and one return bend. Calculate the total head loss across the pipe system. Again, assume the oil used is that of Example 2.3.

Solution: Add the effects of the laminar flow in the pipe to the effects of the head loss in the fittings. Calculation of velocity and Reynolds number from Equations 3.2 and 3.12c

$$v = 0.408\frac{Q}{D^2} = 0.408\frac{(6)}{1} = 2.45 \text{ ft/s}$$

From Equation 3.12c

$$\text{Re} = \frac{7740(2.45)}{11.8} = 1605$$

This value of Re (less than 2000) confirms that laminar flow is actually occurring.

Head loss due to pipe flow, from Equation 3.16

$$H_L = \frac{64}{\text{Re}} \times \frac{L}{D} \times \frac{v^2}{2g}$$

Head loss due to fittings is the sum of each of the terms

$$H_F = K_L \frac{v^2}{2g}$$

$$\text{Total head loss} = \underset{\substack{\uparrow \\ \text{laminar pipe} \\ \text{loss}}}{\frac{64}{1605} \times \frac{100}{1} \times \frac{12(2.45)^2}{2(32.2)}} + [(2 \times 0.19) + \underset{\substack{\uparrow \\ \text{gate} \\ \text{valves}}}{(4 \times 0.8)} + \underset{\substack{\uparrow \\ 90° \\ \text{elbows}}}{(1 \times 2.2)]}\underset{\substack{\uparrow \\ \text{return} \\ \text{bend}}}{\frac{2.45^2}{2(32.2)}}$$

$$= \left[1200\left(\frac{64}{1605}\right) + 5.78\right]\frac{2.45^2}{2(32.2)}$$

So total head loss $= H_L + H_F = 5.0$ ft of oil ◀

Summary

From the elementary concepts of mechanics given in Chapter 2, it has been possible to give properties of fluids in motion. Flow continuity conditions, the energy equation, and power required to generate a fluid flow have been given. The concept of fluid viscosity has also been developed. It is of great importance to be able to calculate what pressure is required to provide a given flow rate through a pipe system. It is possible from the expressions given in this chapter to do this in the case of a pipe system in which laminar or turbulent flow is occurring.

Questions and Problems

1. Write in your own words what the equation of continuity means.
2. If the mean fluid velocity in a pipe is 12 ft/s, what is the rate of flow in gpm? (a) in a 4 in. diameter pipe (b) in a 1½ in. diameter pipe.

3. Fluid flows in a pipe at a rate of 500 gpm. The pipe reduces its diameter at a given point. If the fluid velocity is to be a maximum of 13 ft/s, calculate the minimum value of the reduced diameter.

4. You know that 0.2 slugs/s of MIL-H-5606B hydraulic oil is flowing in a pipe system. Compute the velocity in the pipe where the diameter is (a) 1.25 in. (b) 3 in.

5. What does the Bernoulli equation indicate?

6. Oil flows in a horizontal pipe line of 1½ in. diameter at the rate of 150 gpm. The diameter increases to 3 in. If the pressure in the 1½ in. diameter portion is 1500 psi, calculate the pressure in the 3 in. diameter portion, assuming no loss in energy between the two points considered. Take sg = 0.87.

7. Details of water flow in a pipe at two different points are

	Diameter of pipe	Distance above sea level (ft)
Point A	3 in.	925
Point B	2 in.	937

If the flow rate is 500 gpm, what is the pressure difference between A and B, assuming no pipe friction? Is the pressure difference a loss or a gain?

8. A sharp edged orifice of 1 in. diameter is mounted in a pipe of diameter 3 in. The pressure drop across the orifice is 110 psi for a flow rate of 215 gpm. If the sg of the fluid flowing is 0.85, calculate the orifice discharge coefficient.

9. Hydraulic oil MIL-H-5606B is being driven through a Venturi meter with a throat diameter of 0.75 in. and an entrance diameter of 1.35 in. If the pressure registered at the throat is 650 psi, calculate the pressure at the entrance, assuming that C_d for the Venturi is 0.96, and the flow rate is 70 gpm.

10. Aircraft phosphate ester is being pumped through a pipeline of 1.2 in. diameter at a speed of 3.5 ft/s at (a) 0° F (b) 210° F. What is the Reynolds number at each temperature? Will the flow be laminar or turbulent?

11. Compute the range of mean velocity in the pipe in Problem 10 for the transition Reynolds range if the temperature of the fluid is now 100° F.

12. A pipe diameter is 0.5 in. and 225 ft in length. If oil of viscosity 8 cP and an sg of 0.85 is pumped through it at the rate of 1.5 gpm, calculate the head loss through the system in feet of oil. If the outlet end is now raised by 5 ft, calculate the pressure required to drive the oil at the same flow rate.

13. The value of friction factor for a given pipe condition is 0.025 and the flow is turbulent. Using the Blasius approximation, figure the pipe diameter if the fluid of kinematic viscosity 11 cS is flowing at 28.5 ft/s.

14. In Problem 13, calculate the mass flow rate and the pressure loss in 85 ft of pipe. Assume $\gamma = 53$ lb/ft.^{3}

15. Figure 3.7 shows a hydraulic system. If the flow rate through the system is 50 gpm and the fluid is MIL-H-5606B at a temperature of 100° F, calculate the pressure required from the pump to drive the fluid around the circuit. Both gate valves are fully open, and K_L for each of the four elbows may be assumed to be 0.8 (based on velocity in smaller pipe).

16. Prove to your own satisfaction that Equation 3.18 is correct.

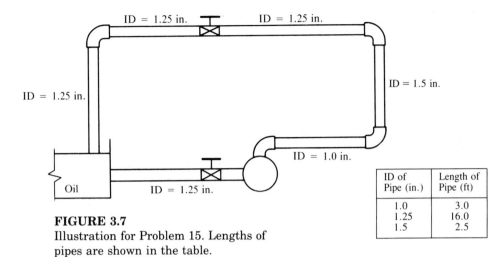

FIGURE 3.7
Illustration for Problem 15. Lengths of
pipes are shown in the table.

ID of Pipe (in.)	Length of Pipe (ft)
1.0	3.0
1.25	16.0
1.5	2.5

Suggested Student Activities

1. *Conversion of pressure into velocity energy.* Measure the pressure generated in your lab faucet. Turn the faucet wide open and measure the flow rate by timing how much water accumulates in a vessel in a given time. Measure the approximate water jet diameter and determine a value of v, the mean water velocity. See if you can calculate the ratio of the pressure of the water main to the quantity $v^2/2g$. Comment on your results.
2. *Reynolds number estimate.* Run the faucet slowly so that the edges of the vertical jet are smooth. Gradually increase the flow until the edges become ragged. Again measure the flow rate and obtain the value of jet mean velocity. Figure the Reynolds number at the point where the different flow rate is obtained using $\mu_{water} = 1$ cP.

Suggested Readings

Esposito, Anthony. *Fluid Power with Applications.* Englewood Cliffs, N.J.: Prentice-Hall, 1980.

Sullivan, James A. *Fluid Power: Theory and Applications,* 2nd ed. Reston, Va.: Reston, 1982

Wolansky, William; Nagohosian, John; and Henke, Russell W. *Fundamentals of Fluid Power.* Boston: Houghton Mifflin, 1977.

CHAPTER 4 ▬▬▬▬

Storage and Distribution Systems

Introduction

In this chapter, useful definitions, reservoirs, conductors, and fluid power seals are discussed. The proper design, selection, installation, and maintenance of fluid power storage and distribution systems are critical to efficient and continued operation of many precision hydraulic components. After completing this chapter, you will be able to define the key terms listed at the beginning of this chapter, name the major functions of a hydraulic reservoir, and calculate the velocity of fluid in a given size conductor.

Key Terms

Additive: A chemical added to a fluid to change its properties.

Conductor: A system component whose primary function is to contain and direct fluid.

Connector: A device for joining a conductor to a component part or to other connectors.

Contaminant: Detrimental matter in a fluid.

Emulsion: A liquid mixture in which one substance is suspended in minute globules within another (e.g., water in oil).

Filter: A device whose primary function is the retention by a porous medium of insoluble contaminants from a fluid.

Heat Exchanger: A device that transfers heat through a conducting wall from one fluid to another. Includes both coolers and heaters.

Manifold: A conductor that provides multiple connection parts.

Micron: A unit of length equal to one-millionth meter, or 0.000039 in.

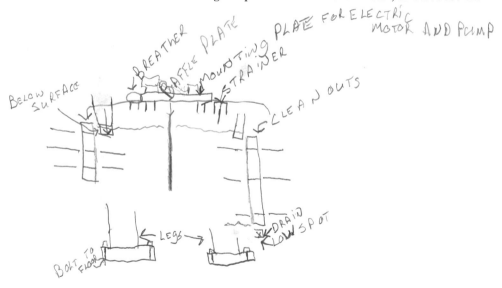

4.1 RESERVOIR AND CONDUCTING NETWORK

Reservoir, conductors, and fittings are essential in all operative hydraulic systems. While the major purpose of a reservoir is to store hydraulic fluid, a properly designed reservoir also traps contaminants, separates entrained air from the oil, and cools the hydraulic fluid.

In stationary, industrial applications where space is not a limiting factor, a remote reservoir is often an appropriate choice (Figure 4.1). The reservoir can be placed where it is easily maintained and accessible for inspection. Sometimes it can be located away from contaminated or excessively hot areas.

On the other hand, relatively small, compact, or mobile equipment may have to include an integral reservoir built into the tubular frame or structure design of the machines. Again, it is important to give consideration to access for maintenance, cooling, and permitting the escape of entrapped air.

A reservoir is more than a simple storage vessel. The volumetric capacity of a reservoir will vary from one-and-one-half to three times the pump output per minute. It must be large enough to hold all the fluid when the actuators are retracted, and allow space for expansion, so foaming will not occur. The larger the reservoir, the more metal surface area is available to transfer heat to the surroundings and cool the fluid.

The return line (Figure 4.2) illustrates that the lower end is submerged below the level of the fluid to prevent aeration, and placed so that the return oil impinges against the reservoir wall to cause fluid circulation and cooling effects.

The inlet pump line is equipped with a strainer resting slightly above the bottom to avoid picking up settled contaminants on the bottom of the reservoir. Increasingly, more hydraulic systems have a pump inlet line equipped with a filter rather than a strainer. This pipe has to be large to avoid starving the pump, which would cause cavitation and admit air into the oil stream.

Figures 4.2 and 4.3 illustrate the use of an angled 100 mesh screen. This screen reduces the amount of entrained air in the reservoir. The use of multiple baffles to extend the distance of flow of the fluid from the return line to the inlet line permits the settling of the contaminants as well as cooling of the fluid. In typical welded reservoirs for industrial applications, a cleanout plate on both ends makes access for cleaning relatively easy. A level indicator enables maintenance personnel to scan the level of fluid visually during operating conditions to see that it is within the critical range.

Since the level of the hydraulic fluid varies with the demands and cycles of the operational components, the ambient air above the fluid must be allowed

Hydraulic reservoir plate

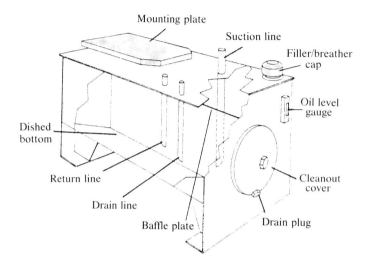

FIGURE 4.1
The major parts of a reservoir: (a) Basic design of a hydraulic reservoir. (b) Industrial hydraulic power unit. (Courtesy The Oilgear Company)

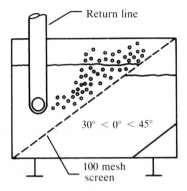

FIGURE 4.2
Screen helps remove entrained air in return fluid.

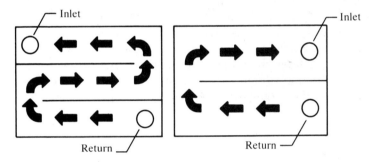

FIGURE 4.3
Installation of baffle walls permits longer settling flow period for contaminants in returning fluid.

to escape and enter the reservoir. For this reason the air breather/filter must be clean to allow air to move freely.

The filler opening should be equipped with a screen to capture or strain any foreign matter and prevent it from entering the reservoir. It should be ensured that clean fluid is properly stored and added when needed. This step is often overlooked. The filling cap should be properly secured.

Well-designed reservoirs are also equipped with a drain plug to allow periodic draining of contaminated fluid and cleaning. Some drain plugs have permanent magnets to attract ferrous particles suspended in the fluid. Seals must also be compatible with the particular hydraulic fluid. Maintaining clean, cool, and consistent fluid in the hydraulic system increases the life of components and reduces the down-time of the hydraulic equipment. This not only increases efficiency but is also more economical, since maintenance costs and down-time are reduced. The source of many hydraulic problems stems from neglect of the fluid in the reservoir. Hydraulic fluids must be compatible with the particular systems operating within given work environments. For example, fire-resistant fluids would be used in environments of elevated temperatures.

4.2 CONDUCTORS

Conductors interconnect the major components in a hydraulic system to confine and conduct the fluid within the system. There are almost endless varieties of manifolds, pipes, hoses, and fittings available to the hydraulic circuit designer. The operation of the hydraulic system depends on the efficiency, serviceability, and leak-free conductors connecting the components within the system. These conductors must withstand the system working pressures, including peak shock pressures periodically created within the system. Conductors must be sized properly for transmitting the required volumetric flow rate of the fluid demanded by any branch of a circuit. Proper design and selection of conductors is influenced primarily by the operating pressures, flow rate, and safety.

Hydraulic lines should be leak-proof and not subject to excessive turbulence or restriction to flow. They should be strong enough to withstand the maximum pressure, temperature, and vibration to which the system may be subjected.

4.3 SELECTION OF CONDUCTORS

A number of considerations are critical in selecting hydraulic conductors for a particular application. These include

- line strong enough to contain the fluid at the calculated working pressure with an adequate safety margin for intermittent surges in pressure,
- lines strong enough to support in-line components that are subject to vibration,
- conductors of the appropriate diameter as determined by the required flow rate to prevent excessive pressure drops,
- lines with a reasonably smooth interior surface to reduce turbulence and frictional losses,
- line material compatible with the hydraulic fluid used in a system to prevent corrosion,
- terminal fittings must be available at all junctions and may require removal for repairs and maintenance, and weight and cost are other factors to be considered.

FIGURE 4.4
Industrial cylinders provide fluid power "muscle" to bend tubing on this automatic bending machine. (Courtesy Aeroquip Corporation)

4.4 PIPE

The American standard pipe and pipe fittings are defined by ANSI standard. The outside diameter of a pipe is held constant for a given nominal size so that threads cut into the outside diameter (OD) must always fit those tapped into a mating port or fitting. The inside diameter (ID) of a pipe will vary according to the standard pipe schedule, which refers to wall thickness. Commonly used pipe in plumbing hydraulic circuits includes ten pipe schedules: 40, or standard pipe; 80, or extra heavy; and 160, or double extra heavy. Pipes are tested for working pressure, which designers use to match systems' pressures with a margin of safety, and burst pressure, at which a pipe theoretically bursts. The thick wall pipes are formed by hydraulically equipped bending machines (Figure 4.4).

4.5 SIZING PIPES

The inside diameter of a conductor is a critical dimension in determining the flow rate capacity of that particular size of conductor (Figure 4.5).

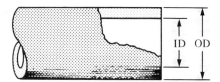

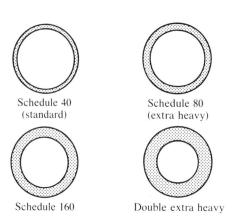

Schedule 40
(standard)

Schedule 80
(extra heavy)

Schedule 160

Double extra heavy

FIGURE 4.5
The OD (outside diameter) remains the same for any schedule pipe, but the ID (inside diameter) decreases as wall thickness increases.

It is possible to compute the inside diameter of the conductor if the required flow rate is given by using the following formulas.

$$\text{area (in.}^2) = \frac{\text{flow rate (gpm)} \times 0.3208}{\text{velocity (ft/s)}}$$

or in SI units

$$\text{area (cm}^2) = \frac{\text{flow rate (liters/min)} \times 0.167}{\text{velocity (m/s)}}$$

For example, if the system required a 20 ft/s velocity in the pressure lines at the flow rate of 10 gpm, the required area of the pipe would be

$$A = \frac{10 \times 0.3208}{20} = 0.1604$$

$$A = \frac{\pi D^2}{4}$$

$$A = 0.7854 D^2$$

$$D^2 = \frac{0.1604}{0.7854}$$

$$D^2 = 0.2042$$

$$D = 0.4519 \text{ in.}$$

Nominal size pipes actually do not exist. The inside diameter (ID) is close but not equivalent to the nominal size. As may be observed in Figure 4.6, a pipe of a nominal size of 1 in. has an actual inside diameter of 1.049 in. for schedule 40 or standard pipe, and only 0.957 in. for double extra heavy pipe. As the walls of a pipe increase and the OD remains constant, the inside diameter decreases by double the difference.

Proper sizing, least restrictive fittings, proper mounting, and leak-free joints will contribute to a more efficient piping installation and system operation.

Nominal Size, In.		ID Wall Thickness			
	OD	Schedule 40	Schedule 80	Schedule 40	Schedule 80
1/8	0.405	0.269	0.215	0.068	0.095
1/4	0.540	0.364	0.302	0.088	0.119
3/8	0.675	0.493	0.423	0.091	0.126
1/2	0.840	0.622	0.546	0.109	0.147
3/4	1.050	0.824	0.742	0.113	0.154
1	1.315	1.049	0.957	0.133	0.179
1¼	1.660	1.380	1.278	0.140	0.191
1½	1.900	1.610	1.500	0.145	0.200
2	2.375	2.067	1.939	0.154	0.218

FIGURE 4.6
Schedule designations in terms of inside diameter.

FIGURE 4.7
Pipe reducer.

Pipe-size reductions, numerous elbows, and poor thread seals generally contribute to undue pressure losses, premature contamination, and hazardous oil leaks. Therefore, it is necessary to select proper pipe size, pressure rating, fluid compatibility, and installation practices. Maintaining cleanliness during installation is very critical to maintenance enhancement.

4.6 TUBING

Rigid hydraulic tubing is designated by its outside diameter. It is generally available in fractional increments of 1/16 in. from 1/8 to 3/8 in. OD; and 1/8 in. increments from 3/8 in. to 1 in. OD; and above 1 in., 1 1/4, 1 1/2, and 2 in. sizes are standard.

Tubing has an advantage over threaded pipe: it can be easily bent, comes in more sizes and materials, and requires fewer pieces and fittings. Tubing can also be cut, flared, and fitted to make a neater conductor system with smoother flow and less chance of leakage.

For higher working pressures, alloy tubing is available in various grades and different wall thicknesses suitable for bending and flaring. For example,

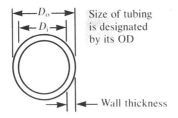

Size of tubing is designated by its OD

Wall thickness

Hose Size	OD Tube Size (In.)	Single-Wire Braid			Double-Wire Braid		
		Hose ID (In.)	Hose OD (In.)	Minimum Bend Radius (In.)	Hose ID (In.)	Hose OD (In.)	Minimum Bend Radius (In.)
4	1/4	3/16	33/64	1 15/16	1/4	11/16	4
6	3/8	5/16	43/64	2 3/4	3/8	27/32	5
8	1/2	13/32	49/64	4 5/8	1/2	31/32	7
12	3/4	5/8	1 5/64	6 9/16	3/4	1 1/4	9 1/2
16	1	7/8	1 15/64	7 3/8	1	1 9/16	11
20	1 1/4	1 1/8	1 1/2	9	1 1/4	2	16

FIGURE 4.8
(a) Size of tubing is designated by the outside diameter. (b) Tubular dimensions.

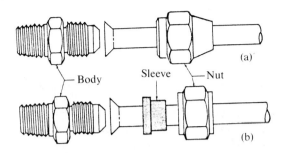

FIGURE 4.9
SAE 37-degree flared tube fitting and self-flaring tube fitting. (Courtesy
Hydraulics & Pneumatics)

AIS 4130 and 8630 are frequently used for high-pressure applications.

Tubing wall thickness can be any dimension ranging from thin to thick. However, the most common dimensions are included in Figure 4.10. The choice of the wall thickness is based on a combination of needed strength and fluid flow capacity.

Two equations that are helpful in making this choice are

$$p = \frac{St}{D_o}$$

where p = maximum working pressure, psi
S = allowable metal stress, psi
t = the wall thickness, in.
D_o = the tube outside diameter, in.

Computing

$$Q = 3.12 \times vA$$

Nominal Diameter, In. (OD)	ID = OD − 2t, where t = wall thicknesses, in.
Available in wide variety, typically, 1/8, 3/16, 1/4, 5/16, 3/8, 1/2, 5/8, 3/4, 7/8, 1, 1 1/4, 1 1/2, and 2 in. in OD (also in metric, typically from 4 mm to 50 mm)	Dozens of ID dimensions are available, from 0.055 in. (1/8 in. tube, 0.035 in. wall) to 1.870 in. (2 in. tube, 0.065 in. wall). Typical values for W: 0.020, 0.028, 0.035, 0.049, 0.058, 0.065, 0.072, 0.083, 0.095, 0.109, 0.120, 0.134, 0.148, 0.165, 0.180, 0.188, 0.203, 0.220, 0.238, 0.250, and 0.259 in.

FIGURE 4.10
Steel tubing dimensions.

where Q = flow, gpm
v = allowable flow velocity, ft/s
A = internal flow area, in.^{2}.

To compute the area, $A = D_i^2/4$. The constant 3.12 converts Q to gpm units. Considerations of safety, pressure loss, temperature, and external mechanical stresses also influence the selection of tubing wall thickness. As a general rule, designers consider one-sixth of the actual tensile strength with a flow velocity of no more than 15 ft/s as acceptable for typical applications.

Tubing has several advantages over other conductors

- Bending quality. Tubing can be easily formed with simple tools.
- Strength. Relatively strong, as no sections are weakened by thread forms as in pipe or crimping as is done in hose fabrication. Specific alloys can provide added strength.
- Weight. Wall thickness is lighter than pipe, thus reducing dead weight.
- Space economy. Bending qualities allow smaller outside diameters and permit more lines in limited space.
- Flexibility. Being less rigid and more continuous tends to reduce vibration.
- Ease of assembly. Bending tubing reduces the need for elbows and other joints, thus reducing possibilities of oil leaks.
- Lower pressure drop. Streamlined flow passage of a tubing system reduces turbulence and larger pressure drops.
- Ease of maintenance. Lower cost of assembly and disassembly, and all other advantages listed makes tubing a favored conductor of hydraulics for many applications.

Tubing is generally assembled with particular attention to the radii of the bends, alignments, and anchoring. Gradual bends tend to eliminate strain by absorbing vibrations and compensate for expansion and construction. Eliminating strain also prevents crystallization and failure of tubing.

4.7 RUBBER AND PLASTIC HOSE

Hydraulic hose is available in a wide variety of rubber and thermo-plastics. A variety of rubber-like elastomers such as neoprene and nitrile rubber are used with various reinforcement materials.

Rubber hose tends to be more tolerant of intermittent high temperatures. Rubber is a thermoset* and therefore does not melt, but it does degrade grad-

---

*That is, rubber is capable of becoming permanently rigid when heated or cured.

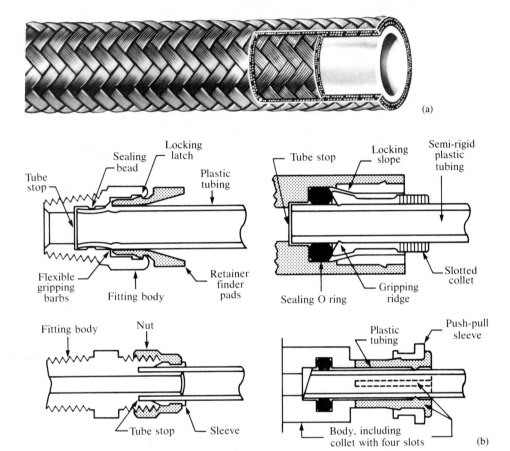

FIGURE 4.11
Hydraulic hoses: (a) Reinforced rubber hose. (Courtesy Aeroquip Corporation)
(b) Hose connections.

ually. However, rubber is heavy, has less tolerance to certain chemicals, swells in oil, and has higher electrical conductivity. Plastic hose is much lighter than rubber hose for the same size. It is also inert to most chemicals and fluids, and can be extruded in continuous lengths. Some plastics, such as Teflon, are compatible with high temperatures up to 450° F. Many combinations of materials (plastics, braided fibers, thermal insulation, and scuff-resistant jacket) are combined for specific user applications (Figure 4.11a).

Unlike tubing, whose dimensions are specified by its OD, hose is designated by its ID. Hence, an −08 size hose means $8/16$, or $1/2$ in. ID.

Hose is commonly used because it is easily installed, will absorb shock, requires less installation skill than pipe or rigid tubing, and is available in a wide range of pressure ratings. Hoses are frequently applied where articulated segments of a machine reciprocate or rotate in relation to each other. Consid-

Leakage	Loss in One Year	
	Barrel (55 Gal)	Vallue, Dollars
One drop in 10 s	0.72	59.40
One drop in 5 s	1.44	118.80
One drop per s	7.44	613.80

The value in dollars of the leaked oil (column 3) is based on a cost of $1.50 per gal. In terms of a machine holding 20 gal of hydraulic oil, the volume losses (column 2) translate into make-up as follows:

One drop in 10 s = 198% make-up per year
One drop in 5 s = 396% make-up per year
One drop per s = 2,093% make-up per year

FIGURE 4.12
Losses by oil leaks. (Courtesy The Rexroth Corporation)

erations in selecting the correct hose for a particular use include the flexing of hoses, constant or frequent flow of fluids, and type of hydraulic fluids. Fittings and quick disconnects are also important considerations for installation and maintenance.

4.8 FITTINGS AND END CONNECTIONS

The most common problem in hydraulic applications is leakage from conductor connections, port connections, or gasketing joints (Figure 4.12). Such leaks can be caused by a lack of suitable joint or gasket material; appropriate tolerances, finishes, thread forms, flares, or swaging; care in forming or machining the joint; proper installation in a particular application; severe system conditions exceeding limits of fittings applied; and inadequate maintenance. The best way to prevent leaks is eliminate separable connections wherever practical. Use of subplates (Figure 4.13) and manifolds can also reduce leaks if connections are made properly. Minimizing the number of piping connections and making all connections accessible for maintenance will also reduce unnecessary external leakage.

4.9 CONNECTORS

The selection of connectors is dictated by port designs on hydraulic components. Ease of assembly, conditions and work requirements, availability of sealants, gaskets, flanges, and other mating surfaces are considerations. There is currently more acceptance of the SAE Straight Thread form fittings in pref-

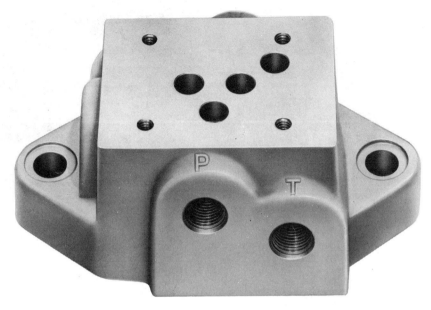

FIGURE 4.13
Use of subplate for component assembly. (Courtesy Miller Fluid Power)

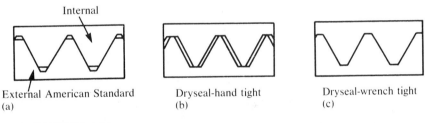

Internal

External American Standard
(a)

Dryseal-hand tight
(b)

Dryseal-wrench tight
(c)

FIGURE 4.14
Pipe thread forms.

erence to the National Pipe Thread form to reduce leakage. The use of polyter-afluoroethylene (PTFE) thread sealant tape or ribbon is widely used on male pipe threads. It is important to wrap the pipe starting about two threads back so portions do not get severed and enter into the system. SAE straight-thread O-ring port fitting also provides a very effective seal. Some manufacturers have converted to SAE straight-thread ports in pumps, valves, subplates, filters, cylinders, and fittings.

There are numerous configurations, sizes, and fastening methods for connectors commonly used in hydraulics. Coatings, pastes, tapes, and design modifications can be used selectively to reduce leakage.

The double-valve quick disconnect hose couplings enable you to join two parts of a hydraulic system about as easily as inserting an appliance cord into

69

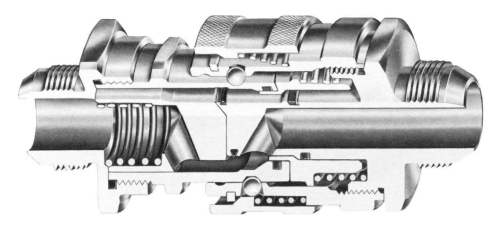

FIGURE 4.15
Quick disconnect coupler. (Courtesy Snap-tite Incorporated)

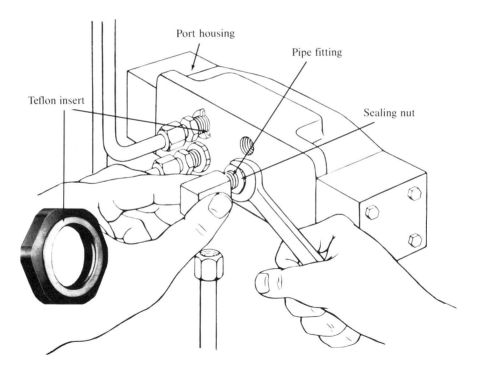

FIGURE 4.16
Teflon seal application. (Courtesy Miller Fluid Power)

an electrical outlet. Connected couplers automatically open the valves, where-as disconnected couplers automatically close the valves. Contamination is a problem with quick-disconnect couplers.

4.10 NEW DEVELOPMENTS IN SEALS

It is estimated that a typical manufacturing plant annually leaks or discards four times more oil than its machines actually hold! Obviously, both good design and improved seals will help reduce the waste and hazards of wasted fluid.

Increasingly, a wider range of plastic seals are becoming available. The polyurethane elastomers provide a combination of hardness, resilience, load-bearing capacity, and exceptional abrasion resistance. The temperature rating of 200° F is moderate. Some of the harder polyurethanes also have a fairly low coefficient of friction.

Teflon seals are suitable for more elevated temperatures up to 600° F and they are self-lubricating. However, they cannot be formed by as many methods as the polyurethanes. Teflon can also be mixed with graphite, glass fibers, metal oxides, bronze, and other strengtheners to provide a wide range of properties.

The most recent technology in sealing materials is creating dual material seal where the desirable characteristics of both materials satisfy the requirements of the seal. One inner O-ring may be made of resilient elastomer, while the outer seal provides the necessary hardness in contact with the shaft.

More recently, the General Electric Silicone Department has created a silicone paste that is spread along the joint and allowed to vulcanize in place at room temperature. The advantage of this type seal is that the silicone is forced into every crevice as the joint is closed before the silicone paste is cured. The surfaces do not have to be prepared as critically. However, the paste does cure relatively quickly. The design and composition of seals must be compatible with the use, conditions, and fluid used in the hydraulic system. Friction, temperature, pressures, and composition of fluid all affect the choice of seals. There is very much current activity in developing new and improved seals. Manufacturer's specifications are a good source of information in guiding the choice of a particular seal for a given application. Periodic replacement of seals is necessary as the material becomes abraded, hardened, or deteriorated.

Seals help keep contamination out, prevent leaks, decrease down-time, and extend the efficient use of equipment. Periodic replacement of seals can prevent costly maintenance of major components. Causes of seal failure can be traced to grit induced into the system, worn shafts, damaged bearings, excessive heat, and failure of seal materials. For more advanced information on seals, see Chapter 19, pp. 495–527 in Yeaple: *Fluid Power Design Handbook,* published by Marcel Dekker, Inc.

Summary

Clean fluid in a hydraulic system requires proper handling of the fluid, sound installation practices, and proper maintenance procedures. The reservoir is more than a storage vessel. It is designed to trap contaminants, separate entrained air from the oil or other fluid, and maintain proper fluid temperature.

The source of many hydraulic problems stems from neglect of the fluid in the reservoir, such as the presence of abrasive contaminants, entrained air, and insufficient fluid to retain working temperature. Proper filtration will prolong the life of precision components in the hydraulic system.

Conductors that interconnect major components are also critical in reducing unnecessary pressure losses, leaks, and other safety hazards. Proper selection and sizing of conductors for a given flow rate will reduce line pressure losses, which are transferred into heat losses.

Seals are also important in preventing leaks, contamination, or safety hazards. A wide range of seals is available, depending upon the particular application and their compatibility with the fluid and adjoining surfaces. Reducing the number of connections and using proper materials and installation practices can greatly reduce oil leaks, which are costly, unsightly, and hazardous.

Questions and Problems

1. What are the primary functions of a hydraulic reservoir?
2. Why is cleanliness important in replacing hydraulic fluid in a reservoir?
3. What purposes do baffles serve in a reservoir?
4. What three types of conductors are used to interconnect hydraulic components?
5. Why is steel tubing more widely used than steel pipe as a conductor of fluids?
6. Compute the diameter of a pipe necessary to conduct the volumetric flow of 5 gpm at 20 ft/s.
7. What is the difference between a flared fitting and a compression fitting for tubing?
8. When would flexible hoses be used in preference to tubing or pipe for conducting fluid?
9. What is the advantage of using quick-disconnect coupling hoses?
10. List four considerations in selecting seals for a hydraulic component?

Suggesting Learning Activities

1. Examine a number of hydraulic system applications and identify the variety of conductors used to transmit fluid flow.
2. Disconnect a pipe, tube fitting, and hose. Be cautious about disconnecting a conductor with pressurized fluid. Reassemble the connectors and check for a leak-proof connection.
3. Disassemble a directional control valve and examine the condition of all seals. Remove and clean all seals and component parts and reassemble. Check the valve for leaks prior to disassembly and after assembly.

4. Write a one-page article on new seal materials in high-pressure hydraulic applications.
5. Visit a plant with hydraulically operated equipment and examine the degree of cleanliness around the reservoir area.
6. Check the fluid level in each of the reservoirs within the fluid power laboratory.

Suggested Readings

Esposito, Anthony. *Fluid Power with Applications.* Englewood Cliffs, N.J.: Prentice-Hall, 1980, pp. 96–113.

Henke, Russell W. "Fluid Power Systems and Circuits." *Hydraulics and Pneumatics Magazine.* Cleveland, Oh., 1983, pp. 285–307.

Seals for Profit Making Designs, Penton 11PC, Educational Division, Rockefeller Bldg., 614 Superior Ave., West Cleveland, Ohio, 44113, 1980. pp. 1–32.

Sullivan, James A. *Fluid Power: Theory and Applications,* 2nd ed. Reston, Va.: Reston, 1982.

Wolansky, William; Nagohosian, John; and Henke, Russell W. *Fundamentals of Fluid Power.* Boston: Houghton Mifflin, 1977, pp. 83–85.

Yeaple, Frank. *Fluid Power Design Handbook.* New York: Dekker, 1984, pp. 451–94.

CHAPTER 5 ▮▮▮▮▮▮▮

Fluid Contamination Control and Fluid Conditioning

Introduction

In this chapter, types of fluid, filters, and fluid conditioning are discussed. You will become particularly aware of the importance of maintaining clean fluids in the hydraulic systems to prolong the efficiency and reliability of a functioning hydraulic system. After completing this chapter, you will be able to describe the different types of hydraulic fluids, the causes of contamination, explain how filters and strainers function, and know how fluid conditioners help retain the inherent qualities of a given hydraulic fluid.

Key Terms

Contaminant: A harmful matter or particulate in a fluid. A particulate may consist of tiny, separate particles as small as bacteria.

Cooler: A heat exchanger that removes heat from a fluid.

Emulsion: A liquid mixture in which one substance is suspended in minute globules within another (e.g., water in oil).

Filter: A device whose primary function is the removal and retention by a porous medium of insoluble contaminants from a fluid.

Filter Media: The porous materials that perform the actual process of filtration.

Heat Exchanger: A device that transfers heat through a conducting wall from one fluid to another. Includes both coolers and heaters.

Hydrolysis: A chemical reaction in which a compound reacts with ions of water to produce a weak acid, a weak base, or both.

Inhibitor: Any substance that slows or prevents undesirable deterioration.

Micron: A unit of length equal to one-millionth meter, or 0.000039 in.

Proportional Flow: A portion of the fluid passing through the filter element while the remaining portion bypasses the filter element.

Strainer: A device that filters coarse particulate.

5.1 HYDRAULIC FLUIDS

Hydraulic fluids transfer energy in fluid power systems. They also serve several other important functions. These include sealing close-tolerance parts against leakage; providing lubricity to minimize wear and friction between adjacent surfaces; removing heat; flushing away contaminants and protecting machined surfaces against corrosion.

Prior to the discovery of oil, water was used in presses as the medium for transferring energy. However, water had several limitations such as its corrosive effects on metals, its lack of lubricity, and its lower freezing point (32° F, or 0° C).

Many improvements have been made in the quality and range of available hydraulic fluids designed to function in particular environments and meet system requirements. Due to the diversity of hydraulic applications, fluid manufacturers have developed a wide range of fluids with specific properties for particular application systems. Among the more common hydraulic fluids are petroleum-base fluids, fire-resistant fluids, synthetic fluids, phosphate esters, silicones, and other synthetics that meet specific equipment and military standards.

Physical properties of fluids were discussed in Chapter 3. We now review those properties and discuss others that are critical to reliable system performance and extended component life.

Viscosity is one of the most important fluid qualities. Viscosity of oil is the measure of the resistance of oil to flow. Low-viscosity oils flow freely, whereas high viscosity oils flow sluggishly. Oil viscosity decreases as the temperature increases and vice versa. An oil whose viscosity changes relatively little for a given temperature change has a high viscosity index (VI). A hydraulic fluid that is too viscous (thick) usually will cause high pressure drops, sluggish operation, and will contribute to low mechanical efficiency and high power consumption. High-pressure systems and high-precision systems rely on low viscosity fluids that permit efficient operation. It is important to recognize how the temperature of a fluid will vary its viscosity. When the fluid is cold, it is likely to move sluggishly and create pump cavitation and slower actuator response. This is particularly noticeable in sub-zero weather. Fluid suppliers' specifications should first be consulted when replacing a hydraulic fluid.

Stability of a hydraulic fluid is another important property affecting fluid longevity and system performance. Fluid turbulence, thermal changes, oxidation, chemical breakdown, and cavitation all tend to alter the properties of fluid and contribute to fluid degradation.

Lubricity is another important characteristic of hydraulic fluids. The fluid must provide lubrication to all moving parts to reduce wear and prolong the use of component parts. One of the advantages of petroleum fluids is their inherent lubrication qualities.

Anti-wear additives can be compounded with other hydraulic fluids to extend their lubrication qualities. Where severe load-carrying requirements have to be met (as with the case of certain types of motors and pumps), lubricity is particularly important.

Aeration and foaming resistance are other considerations in determining fluid quality. To some extent, fluids contain some amount of air entrained in solution. Excessive entrained air tends to increase fluid compressibility, thereby making the pressurized system noisy, erratic, and elastic. Compression of entrained air creates unnecessary heat and increases oxidation of the fluid. When bubbles collapse on the discharge side of the pump, eventual damage to the pump elements can result. Placing a fine mesh screen in the reservoir will tend to release air from the fluid. Anti-foaming additives also enhance the release of air from the fluid.

Corrosion prevention is another essential characteristic of a quality fluid. Moisture in various degrees seems to be present because of thermal changes in a system and resulting condensation or hydrolysis. Hydrolysis is a chemical reaction in which a compound reacts with ions of water to produce a weak acid, a weak base, or both. Since most hydraulic components are made of ferrous materials, corrosion prevention is necessary. The detrimental effect of corrosion is that the tolerances of close-fitted parts increases internal leakage. Rust inhibitors added to the oil or other fluids tend to create a surface film, which reduces the corrosive effects within the components.

Materials compatibility is also important in the selection and use of a particular fluid. The fluid must be compatible with the seals. Petroleum-based fluid tends to swell natural rubber seals, but is more compatible with nitrile rubber seals. Teflon seals, for example, are very inert and do not deteriorate as readily due to chemical reaction with the fluid. Compatibility of fluids also extends to acceptance of such additives as detergents, reaction to metals in suspension or solution, filter medium, fire resistance, and paints and elastomers.

5.2 TYPES OF FLUIDS

Petroleum oils are the most common hydraulic fluids used in a wide-range of hydraulic applications. The characteristics of petroleum-based hydraulic oils are controlled by the type of crude oil used (naphthenic, aromatic, or paraffinic), the method of refining, and the additives included.

Naphthenic oils have good resistance to emulsification and a high resistance to foaming. Such oils are suitable for systems that work at temperatures as high as 150° F. However, their relatively low viscosity index makes them unsuitable for use in systems where the oil temperature varies too widely.

The aromatics are also a derivative of the hydrocarbons, with a higher presence of benzene, and they are more compatible with moderate temperature variation.

Paraffinic oils have a higher viscosity index than napthenic oils, and they are more suitable for use in systems where the temperature varies greatly. Oils of this type will perform satisfactorily in the presence of high moisture, contaminants, other accelerators, or oxidation. Moderate temperature fluids can be attained from super-refined natural petroleum products consisting of complex mixtures of hydrocarbons in the form of paraffins, naphthenes, and aromatics. The additional refining steps eliminate undesirable molecules and contaminants. Fire-resistant fluids are necessary when a hydraulic system is located near high-temperature equipment or sources of ignition that could endanger personnel, equipment, or facilities. The use of fire-resistant fluids in such areas as mining, steel mills, foundry, die-casting, heat treating, welding, aircraft, space vehicles, and naval vessels is a major consideration in the design and operation of hydraulic systems. Fire-resistant fluids include water, water-glycol, water-oil emulsions, phosphate esters, and chlorinated hydrocarbons.

Water as a hydraulic fluid is highly fire resistant, readily available, and inexpensive, but it has severe limitations. Water contributes to corrosion, evaporation, poor lubricity, freezing, and cavitation. When water is used as a hydraulic fluid, it is often combined with certain oils or glycols to provide the necessary properties to resist fire and rust, and aid lubricity.

Water-glycols also provide excellent fire resistance because they contain between 35 and 55 percent water. Water glycols are solutions rather than emulsions, so they can be formulated for a range of viscosities similar to hydrocarbon oils. It is also possible to include additives that improve viscosity, reduce corrosion, and increase lubricity. Water-based fluids usually are not recommended for the high bearing loads exerted on the ball bearings of axial-piston pumps, or for pressures greater than 1500 to 3000 psi.

Water-in-oil emulsions contain 40 percent water. The rest is oil, emulsifiers, and other additives. The water is dispersed in microscopic droplets surrounded by a film of oil. Because the oil is the outside layer, lubricity is improved compared to the oil-in-water emulsion. This fluid has many of the same properties as petroleum oil, yet it is fire-resistant due to the high water content. This emulsion has a high shear rate and therefore flows readily through high shear orifices such as those found in pumps and valves.

Oil-in-water emulsions contain about 95 percent water and 5 percent soluble or emulsible oil and additives. Such additives may include emulsifiers, wear inhibitors, oxidation inhibitors, vapor inhibitors, and bactericides.

In this emulsion, the oil is dispersed in fine droplets in the water and each droplet is encased in a continuous layer of water. While fire resistance is high, lubricity is greatly diminished because of the high water content. The emulsifier controls the degree of oil dispersion, the stability, and the size of the coated droplets.

Phosphate esters are also fire resistant, but not to the same degree as water-glycols or water-oil emulsions. Phosphate ester is an organic alcohol attached to a phosphorous atom. The lubricity of phosphate esters is equivalent to petroleum oils and has high thermal stability. Because phosphate ester fluids are strong solvents for many plastics and elastomers, seals made of butyl rubber, ethylene-propylene, or fluoroelastomer are compatible with this fluid. Phosphate-ester fluid serves as an excellent detergent and prevents build-up of sludge on the hydraulic components.

Silicone fluids have excellent thermal stability. This means they resist physical and chemical change under severe heat, cold, shear, oxidation, and other operating conditions that normally break down organic fluids. Silicone fluids are generally dimethyl polysiloxanes. This synthetic fluid is inert, non-corrosive, non-toxic, and has low volatility. These fluids are compatible with nitrile rubber, fluorel, and teflon suspensoid seals. Fluids of the future will likely continue to concentrate on synthetic solutions compounded to provide desirable characteristics that match system requirements and environmental, safety, and stability concerns over a wide range of thermal parameters.

When filtration is improved, precision of components is improved and performance specifications are enhanced. Fluid specifications will also have to be raised and matched with specific applications. The important considerations are to follow manufacturer's specifications, keep the system fluid clean with regularly scheduled maintenance, and operate the system within the temperature and pressure limits for which the equipment is designed.

5.3 FILTRATION OF FLUIDS

The primary task of any filter is to remove and retain unwanted and destructive substances from a hydraulic medium. An extremely high percentage of hydraulic system down-time can be attributed directly to fluid contamination. As the pressure ratings in hydraulic systems increase and the tolerance clearances within the components decrease due to precision machining, contamination control of the fluid becomes most critical. Component or system failure causes down-time, increases maintenance costs, and reduces productivity and profits. Some components, particularly valves with small clearances, tend to have a low tolerance for larger particulate contamination in a system and can malfunction easily (see Figure 5.1). Contamination control must be an integral part of any fluid power system.

Ideally, it would be desirable to exclude all contaminants from a hydraulic fluid, but this is virtually impossible. Even though hydraulic oil is refined and blended under relatively controlled conditions, the improper storage of oil,

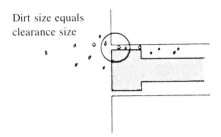

FIGURE 5.1
Contaminant size equals clearance size.

the process of transferring it to the reservoir, and contaminants found in the air as well as in conductors and components are all potential sources of fluid contamination.

5.4 CONDITIONING OF FLUIDS

The storage of new hydraulic oil in a clean area and the transfer of the oil through a filtration apparatus will reduce the amount of contamination when replenishing or replacing oil (Figure 5.2). The reservoir should be drained and cleaned periodically to remove sediments and other accumulated contaminants.

When new or rebuilt components are replaced in a system, a variety of contaminants can also be introduced into the system. Such components should be cleaned and thoroughly flushed as soon as possible after replacement. Conductors (particularly steel) will also corrode and scale, especially with extended nonuse. Another source of contamination is the environment in which a hydraulic system operates. Some environments contain more contaminants

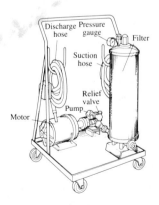

FIGURE 5.2
Replenishing hydraulic fluid using high-capacity element filter.

than others. For example, a hydraulic forklift working in a cement factory, or a hydraulic hoist in a foundry may have severe exposure to contaminants. Therefore, preventative maintenance in such work areas must be done very frequently compared to a relatively dust-free environment.

Since it is not possible to completely eliminate contamination, precautions should be taken to reduce the effects of contamination and to use filtration to remove unwanted particles. Keeping the fluid clean and preventing unnecessary contaminants from reaching numerous sources is the first step toward increasing system reliability. Contamination may also be generated within the system itself. As components and numerous parts function within a system, particles will form. This formation can come from contact wear, erosion, corrosion, moisture from water condensation inside the reservoir, worn seal materials, and sludges due to oxidation of oil build-up. It is necessary to use continuous filtration so that these particles are removed and retained in the filter media, and not in the circulating fluid. Whether the source of contaminants are built-in, added-to, ingressed, or generated, the important thing is to detect the presence of such contaminants and remove them to the level where harmful effects are minimized. Where reliability is a critical factor, such as in an aircraft, good maintenance practices contribute significantly to the overall safety of the hydraulic systems.

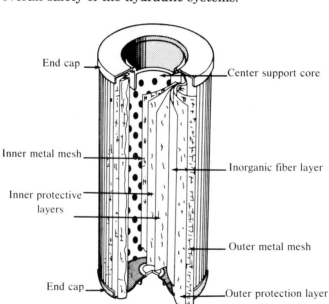

FIGURE 5.3
Hydraulic filter is pleated to increase the effective filtration area. (Courtesy *Hydraulics & Pneumatics*)

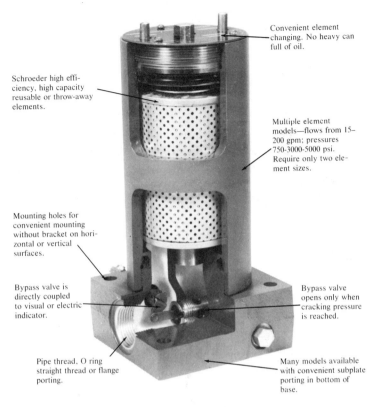

Convenient element changing. No heavy can full of oil.

Schroeder high efficiency, high capacity reusable or throw-away elements.

Multiple element models—flows from 15–200 gpm; pressures 750-3000-5000 psi. Require only two element sizes.

Mounting holes for convenient mounting without bracket on horizontal or vertical surfaces.

Bypass valve is directly coupled to visual or electric indicator.

Bypass valve opens only when cracking pressure is reached.

Pipe thread, O ring straight thread or flange porting.

Many models available with convenient subplate porting in bottom of base.

FIGURE 5.4
Base-ported filter. (Courtesy Schroeder Brothers Corporation)

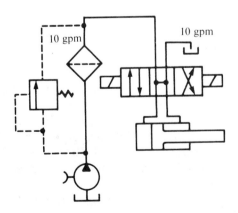

10 gpm

10 gpm

FIGURE 5.5
Full-flow, in-line pressure filtration with bypass valve.

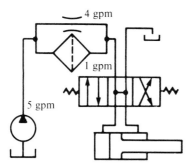

FIGURE 5.6
Proportional flow filtration.

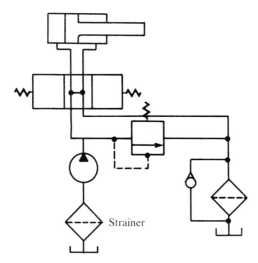

FIGURE 5.7
Basic circuit design for return-line filtration.

The effects of contamination from whatever source may plug the filter element and create a bypass flow, or it may plug an orifice such as in a servo-valve. If a particle's size cannot be tolerated, it may bind the spool of a valve and cause a system failure. More commonly, the presence of cumulative contamination leads to degradation, failure of the fluid, and premature wear process on component parts. This leads to more equipment oil leakage and accelerated harmful effects.

Particle size, distribution, and bypass oil leakage will affect the degradation of a fluid. Hydraulic filters generally contain disposable, high-performance paper cartridges made of inorganic fibers impregnated with epoxy resin. Typically, the element includes end caps that provide a means of mounting the element within its housing. An O-ring is generally used to seal the element

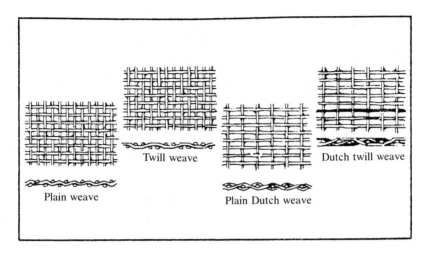

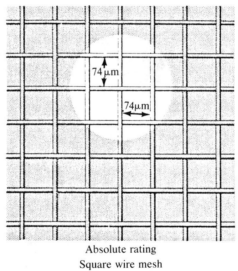

Absolute rating
Square wire mesh

FIGURE 5.8
Basic wire mesh weaves. (Courtesy *Hydraulics & Pneumatics*)

against leakage in the housing. Other mechanical supports that hold the medium shape and resist its collapse are provided to allow fluid to pass through a cartridge readily. All this support hardware should be structurally capable of withstanding the pressure forces in the filter. This includes full-flow by-pass valve pressure differential generated by the pressure surges in a given system. Failure of the element hardware or filter medium defeats the purpose of a filter.

Types of filtration include the full-flow, in which 100 percent of the oil circulated to the hydraulic equipment passes through the filter (see Figure

5.1). This type filter is usually provided with an automatic bypass valve so that if the medium becomes clogged, the hydraulic system will not be interrupted.

Such filters have dirt indicators that reveal the extent of filtration in the operational system. In a full-flow filtration circuit (Figure 5.5), it is important that maintenance personnel change clogged filters prior to the flow by-passing the filter medium.

In certain applications, such as the actuator on an aircraft retractable landing gear, it is critical that the actuator be usable, even if the filter in the circuit is clogged.

Almost all filter housings utilized in hydraulic systems include some type of bypass or relief valve. This device prevents excessive pressure drops across the filter element. Both the bypass valve and the element/housing seal interface offer a very real leakage potential. However, in full-flow filtration circuits, most of the oil is filtered during the operational cycle.

There are applications where only some of the fluid flows through the filter element. The rest will bypass when the system flow exceeds the capacity of the filter (see Figure 5.6). Where a system does not operate continuously, the frequency of contaminants may be sufficiently low so that proportional filtration will be adequate. In some systems, partial return-line filtration (Figure 5.7) is acceptable with the bypass valve (internal or external) always passing a percentage of the flow. Within such a circuit, the effective filtration depends on the continuous rate of flow across the filter element. Depending on conditions, 10–25 percent of the total oil flowing through the filter element may be adequate. However, more manufacturers of equipment would rather build in system reliability by including full-flow filtration.

5.5 MATERIALS USED IN FILTER ELEMENTS

There are three major types of filter element materials: mechanical, absorbent (inactive), and adsorbent (active). Mechanical filters use finely woven, metal-wire screens or discs. Such elements are used more commonly to remove fairly coarse, insoluble particulates. Absorbent (inactive) filter media are constructed of porous, permeable material such as cotton, diatomaceous earth, cloth, paper, fiberglass, and other inorganic fibers. As the fluid permeates the element material, the absorbent action causes the contaminant particles to be trapped by the walls of the medium. Absorbent media elements make it possible to filter extremely small particles and once they become clogged with contaminants, many filters are disposable. Wire mesh elements can be cleaned to required levels following proper procedures and using appropriate solvent cleaners.

Adsorbent (active) filter element materials adhere to the surface of cer-

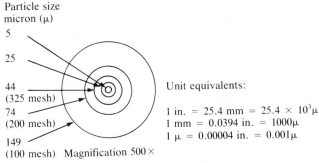

Particle size
micron (μ)

5

25

44
(325 mesh)
74
(200 mesh)
149
(100 mesh)   Magnification 500×

Unit equivalents:

$1 \text{ in.} = 25.4 \text{ mm} = 25.4 \times 10^3 \mu$
$1 \text{ mm} = 0.0394 \text{ in.} = 1000\mu$
$1 \mu = 0.00004 \text{ in.} = 0.001\mu$

Microns diameter

| 0.0001 | 0.001 | 0.01 | 0.1 | 1 | 10 | 100 | 1000 | 10,000 |

1A   Light waves Limit of visibility   1 cm
Atoms   Colloids Lower limit of visibility to the naked eye

Magnification 500×	U.S. and ASTM standard sieve numbers	μ rating	In.
	60	238	0.009
	100	149	0.0055
	200	74	0.0027
	325	40	0.0016
	625	20	0.0008
	1250	10	0.0004

FIGURE 5.9
Micron equivalents.

tain suspended solids or liquids. Adsorption is a surface phenomenon. This means the capacity of a solid to adsorb depends on the extent of its surface exposed to the particles of another substance, as well as its chemical nature. Fuller's Earth, charcoal, activated clay, and chemically treated paper are used in the elements of adsorbent filters. These filter elements remove contaminants mechanically and by the ionic attraction that is part of the adsorbent material.

Strainers, commonly made of wire mesh screens (Figure 5.8) and used on the inlet lines and reservoir fill pipes, are sized by the number of meshes per square inch (60) or the degree of fitness of weave. Filter elements are rated by micron (5, 10, etc.). A micron is a unit of length equal to one-millionth meter. Twenty-five microns equal approximately 0.0001 in. (see Figure 5.9).

There are a number of important considerations in controlling contamination in a hydraulic system:

- Consider the types and extent of possible contaminants.
- Consider the environment that allows contaminants to be introduced into the system.
- Install as fine a filter as necessary.

87

- Protect critical components first.
- Install filters in circuit locations that provide the most protection with appropriate flow rate and capacity.
- Prevent all unnecessary contamination with regularly scheduled maintenance.
- If serious problems arise, do a laboratory analysis of the fluid sample to detect the types, frequency, and sources of contamination.
- Keep good records of service time performed by the equipment, including the filter elements. Remember: the primary purpose of the filter is to remove contaminants from the system fluid. Otherwise, they will accelerate the wear of numerous critical components in the operating system.

5.6 HEAT EXCHANGERS

Temperature regulation of hydraulic fluids is an important consideration in maintaining the narrow limits of viscosity changes and leakages. There are several ways to avoid overheating the fluid. First, conserve energy by pumping only the amount of fluid needed to operate the system. Provide plenty of cooling surface, use higher temperature range fluids, or if necessary add a heat exchanger (Figure 5.10).

Ideally it is better to design a system that can operate without a heat exchanger. Heat loss is an unnecessary waste of energy. If the reservoir and conductors are not able to dissipate the extra heat generated, it may be necessary to include a heat exchanger to maintain a tolerable heat range.

If the fluid gets too hot, it changes its viscosity index, tends to oxidize, and may eventually reduce the lubricating quality. Coolers and heaters are considered heat exchangers, and are used to maintain optimum temperature of the fluid in an operating hydraulic system.

Heaters are used in colder climates, enabling the fluid to reach the necessary temperature and flow freely through the pump inlet and other conductors. Some synthetic fluids change their viscosity very little in colder temperature ranges and do not need heaters.

Heat is generated whenever fluids are throttled through a flow control valve, pressure regulators, relief valves, undersized piping, and clogged filters and by internal leakage, turbulence, and flow friction. Other sources of heat from electric motors, solenoids, and mechanical systems may also add heat to the system. Heavy duty cycle may add considerable heat to a hydraulic system. Therefore, it is important to understand fluid flows, pressures, and temperature conditions to determine what cooling capacity is necessary. Whenever a fluid is transferred under pressure, there is a power requirement from an external or primary source such as an electric motor or an engine to drive the pump.

As you learned in Chapter 2, energy cannot be created or destroyed, and will revert to heat energy.

Since power is defined as energy delivered per unit time, this energy must be absorbed by the oil and its containers, which will both tend to increase in temperature unless heat is dissipated. When all the energy is absorbed by the oil, an approximate increase in temperature may be determined by the following equation

$$\Delta T = \frac{\Delta p}{9339 C_p \gamma}$$

where ΔT = the incremental rise in temperature, °F
Δp = the pressure drop for this rise in temperature, psi
C_p = the specific heat (for oil, 0.5 Btu/lb °F)
γ = the specific weight (for oil, 0.03 lb/in.^{3})

Thus, there is about a 1°F rise in temperature across any hydraulic resistance (orifice, pipe-elbow, etc.) for each 140 psi drop, if the fluid has a petroleum base for which $\gamma = 0.03$ lb/in.^{3}.

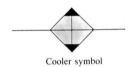
Cooler symbol

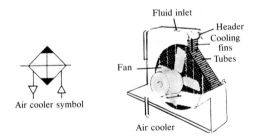
Air cooler symbol

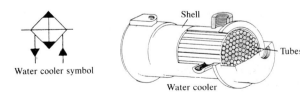
Water cooler symbol

FIGURE 5.10
Heat exchangers maintain operating fluid temperature.

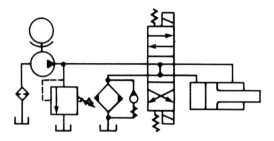

FIGURE 5.11
Heat exchanger in hydraulic circuit.

Example 5.1 ▶ Oil has to be pumped from a reservoir to an accumulator on a skyscraper 700 ft above the pump. Pressure needed in the accumulator is 850 psi. The flow resistance in the pipe is determined to be 215 psi with a flow rate of 65 gpm. For the oil, the specific gravity sg = 0.85, where C_p = 0.45 Btu/lb °F. To compute the increase in temperature (T) of the oil, the following procedures are followed.

Solution: Note that the sources of pressure must be added to obtain the overall pressure required from the pump. These sources include the elevation of 700 ft, the given flow resistance in a pipe, plus the pressure required in the accumulator

- elevation
- pipe resistance
- pressure required in accumulator

$$p = 700 \text{ ft} \times 0.85 \times \frac{62.4 \text{ lb}}{\text{ft}} \times \frac{1 \text{ ft}^2}{144 \text{ in.}^2} + 215 \text{ psi} + 850 \text{ psi}$$

$$p = 257.8 + 215 + 850 = 1323 \text{ psi}$$

$$p = \text{weight of oil per in.}^3$$

$$p = 0.85 \times \frac{62.4 \text{ lb}}{\text{ft}^3} \times \frac{1 \text{ ft}^3}{1728 \text{ in.}^3} = 0.0307 \text{ lb/in.}^3$$

From the previous equation

$$T = \frac{1323}{(9339) \, (0.45) \, (0.0307)}$$

$$T = 10.26° \text{ F}$$

Since some of the heat generated will be absorbed away, this calculated rise in temperature will be the maximum value expected. ◀

COOLERS

Selection of an air-cooled or water-cooled heat exchanger will depend on cost, space available, and the amount of heat to be dissipated. Water coolers are more efficient. Water cooling is recommended for the most compact and efficient operation, but forced air is also widely used. Water-type oil coolers consist of a shell and tubes of various configurations and fins. Water is circulated through the tubes while the hydraulic fluid circulates through the shell, exchanging temperature to the cooler circulating water. The shell of the heat exchanger also helps dissipate heat, further cooling the hydraulic fluid.

Normally, oil coolers are placed on the return line or low pressure side to the reservoir. In air coolers the fluid flows through tubes, while forced air is circulated over the tubes and fins to dissipate some of the heat from the area tubes and fins surfaces.

Overheating oil can cause it to decompose, form varnish on component surfaces, damage seals, cause erratic system performance, increase internal leakage, and lower viscosity. Cold oil tends to be sluggish, causes pump starvation, reduces component response time, and also leads to premature component failure. Both extreme heat and cold have adverse effects on system efficiency and reliability.

Summary

The development of hydraulic fluids from the super-refining of base crudes, the inclusion of many additives to the development of fire-resistant fluids, and most recently the formulation of synthetic fluids have broadened the range and versatility of hydraulic applications in many industrial sectors. It is critical that fluids be compatible with the viscosity, temperature, pressure, and seal requirements of the hydraulic system. Manufacturer's specifications should be consulted for particular equipment, environment and application parameters for the optimum fluid when replenishing or replacing fluid.

The primary purpose of filtration is the removal and retention of contaminants from the system fluid to extend the life cycle of the fluid and all components within the system. An estimated 65 to 86 percent of most hydraulic system problems stem from fluid contamination and poor maintenance practices.

Heat exchangers will likely have less future use as energy conservation becomes more critical and hydraulic system designers find better ways to solve the problems of heat balance within a functioning system. Future synthetic fluids will also become capable of operating within a wider temperature range.

Questions and Problems

1. What are the four functions of a fluid in a hydraulic system?
2. Distinguish the differences between viscosity and viscosity index.
3. Read an article and prepare a one-page report on a new synthetic hydraulic fluid.
4. Name three types of fire-resistant hydraulic fluids.
5. List three major sources of possible fluid contamination.
6. What is the difference between a strainer and a filter?
7. Draw a filter circuit illustrating the placement of a filter in the upstream position relative to the pump.
8. What is the difference between an absorbent and an adsorbent filter element?
9. Explain how a water-cooled oil heat exchanger functions.
10. In Example 5.1, if the pump is 92 percent efficient, calculate the horsepower input to the pump.
11. What are the harmful effects of excessive heat upon the fluid's upper temperature range operating over a prolonged period?

Suggested Learning Activities

1. Visually inspect the air breather cap of the reservoir in your laboratory for contamination.
2. Sketch a cut-away and label the major parts of a filter element and housing assembly.
3. Write to a manufacturer of fire-resistant hydraulic fluids for literature, then write a two-page paper on the composition and characteristics of a particular fluid.
4. Examine any hydraulically equipped machine and locate the oil cooler. Note where it is placed in the circuit and draw the schematic circuit.

Suggested Readings

Basic Hydraulic Filtration, (slides and tape), Parker-Hannifin, Cleveland, Ohio.

Fluid Power Handbook & Directory, Hydraulics and Pneumatics Magazine. Cleveland, Ohio, 1984–1985, Section A, pp. 76–85.

Frankenfield, Tom. *Using Industrial Hydraulics.* Bethlehem, Pa.: Rexroth Corporation, 1979, section 7, pp. 1–19.

Henke, Russell W. "Fluid Power Systems and Circuits." *Hydraulics and Pneumatics Magazine.* Cleveland, Oh., 1983, pp. 195–220.

Wolansky, William; Nagohosian, John; and Henke, Russell W. *Fundamentals of Fluid Power.* Boston: Houghton Mifflin, 1977, pp. 57–78.

Yeaple, Frank. *Fluid Power Design Handbook.* New York: Marcel Dekker, 1984, pp. 424–50.

CHAPTER 6

Energy Input Devices

Introduction

The energy input device used in a high-pressure hydraulic system is called a positive displacement pump. In this type of pump there is a fixed quantity of fluid delivered for a given number of pump shaft rotations. The delivered fluid is at a high pressure. It is supplied to the closed hydraulic circuit where it performs useful work, and then is returned at low pressure to the reservoir. This chapter will give you a thorough understanding of where to use different types of pumps. You should also be able to make design calculations, select a suitable pump from a pump manufacturer's handbook, and work the problems at the end of the chapter.

Cavitation: The condition achieved when the pressure goes below that of the saturated vapor and harmful bubbles come out of solution.

Gear Pump: A pump that relies upon the meshing of gears to drive oil from its present region to the high-pressure region.

Mechanical Efficiency: The power delivered by the pump, expressed as a percentage of the actual power supplied to it.

Overall Efficiency: The percentage of the input power available as direct output power.

Piston Pump: A pump that uses the action of pistons within close fitting cylinders to generate pressure.

Pump: A mechanical device that accepts fluid at low pressure at a given rate and propels it at a higher pressure at the same rate.

Saturated Vapor Pressure: The pressure at which bubbles form in a liquid. These bubbles comprise a mixture of air and the liquid vapor.

Vane Pump: A pump that relies upon contact of vanes with the inner wall of a chamber to generate pressure.

Volumetric Displacement: The quantity of oil delivered at low pressure per revolution of the pump shaft.

Volumetric Efficiency: The percentage of the volumetric displacement achieved at low-delivery pressure that is available at pump-working pressures.

6.1 THE MODE OF ACTION OF A PUMP

Description of Operation. A pump consists of a low-pressure fluid source; a driven, sealed pumping chamber; and a high-pressure outlet. The pumping chamber increases the volume in the inlet region by means of mechanical devices. This creates a partial vacuum, causing the low pressure oil to enter. As the oil moves through the pump its volume tends to be reduced, causing it to be compressed (pressurized) and expelled from the outlet. If the outlet is a reservoir and there is no pressure required to drive the oil, no pressure will develop. However, if the discharge line is blocked or partially blocked, there will be a large resistance to the flow. Because the oil has nowhere to go, it will develop extremely high pressures that could burst pipes and damage components. To summarize, the pressure of delivery is actually dictated by the load in the circuit. To prevent damage, the loading and the pressures must be limited. How to do this will be described in the next chapter.

Operating Properties. The ratio of the power output to the power input is called the overall pump efficiency (e_o). Efficiency is quoted as a percentage, and it will always be less than 100% due to the effects of oil leakage, turbulence, swirl, and mechanical friction. We may write

$$\text{overall pump efficiency, } e_o = \frac{\text{output}}{\text{input}} \times 100\%$$

Thus

$$e_o = \frac{\text{pump output fluid horsepower}}{\text{pump input brake horsepower, Bhp}_{\text{in}}} \times 100$$

From earlier chapters we learned

$$e_o = \frac{\dfrac{(p \times Q_A)}{1.714}}{\dfrac{T_A \times N}{5250}} \times 100 = \frac{p \times Q_A}{T_A \times N} \times 3060 \times 100 \tag{6.1}$$

where T_A = torque on pump shaft, lb·ft
p = pressure increase, thousands of psi

$$Q_A = \text{actual flow rate, gpm}$$
$$N = \text{shaft rotational speed, rpm}$$

Example 6.1 ▶ A pump inlet is situated in a reservoir in the atmosphere and discharges 3.5 gpm when operating at a pressure of 2250 psi. The overall efficiency of the pump is 88%. If the pump is being driven at 1780 rpm, compute the input torque.

Solution: From Equation 6.1

$$e_o = \frac{(2.250)(3.5)/1.714}{\text{Bhp}_{in}} \times 100 = 88\%$$

$$\text{Bhp}_{in} = 5.22$$

The torque input is given by

$$\text{Bhp}_{in} = \frac{T_A \times N}{5250}$$

$$T_A = \frac{\text{Bhp}_{in} \times 5250}{1780} = 15.4 \text{ lb·ft}$$ ◀

6.2 PUMP TYPES

When considering a particular design or a given task requirement, you can choose from three basic design types: gear pumps, vane pumps, or piston pumps. Within these three types we have designs where the volumetric flow can only be varied by changing the rotational speed. These are called fixed displacement machines. Another broad description of "variable displacement" machines can be given those pumps whose elements have physical relationships to each other that can be changed. This brings about a change in flow even though the pump rotational speed remains constant.

The three basic types of pump design discussed earlier are further divided

1. Gear pumps (fixed displacement)
   a. external gear pumps
   b. internal gear pumps
   c. lobe pumps
   d. screw pumps
2. Vane pumps (fixed or variable displacement)
3. Piston pumps (fixed or variable displacement)
   a. axial design
   b. radial design

An external gear pump (Figure 6.1) creates a partial vacuum, thus admitting oil there. Oil is then carried around the outside of the gears. The meshing gear teeth ensure the oil is discharged as shown, since they are enclosed by the pump casing and side plates (also called pressure plates or wear plates). A lobe pump's meshing components are much larger and coarser than those of the gear pump (see Figure 6.2), but the operating principle remains the same.

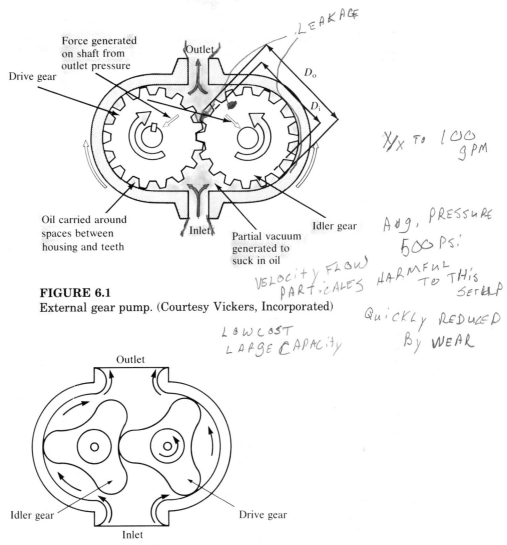

FIGURE 6.1
External gear pump. (Courtesy Vickers, Incorporated)

Handwritten annotations:
LEAKAGE
D_o *D_i*
X/X TO 100 gPM
Avg. PRESSURE 500 PSi
VELOCITY FLOW PARTICALES HARMFUL TO THIS SETUP
QUICKLY REDUCED BY WEAR
LOW COST LARGE CAPACITY

Outlet

Idler gear

Drive gear

Inlet

(Operating principles similar to gear pump)

FIGURE 6.2
Lobe pump. (Courtesy Vickers, Incorporated)

97

To calculate the expected delivery from a gear pump, we calculate the volume in the annular space between the outside and the inside diameters of the gear teeth, D_o and D_i. Thus, we may write V_D, the volumetric displacement, as

$$V_D = \frac{\pi}{4}(D_o^2 - D_i^2) \, L/\text{rev} \qquad (6.2)$$

where L is the width of the gear teeth in in. If D_o and D_i are also in inch units and N, the rate of revolution, is in rpm, the theoretical flow rate, Q_T, is

$$Q_T(\text{in.}^3/\text{min}) = V_D(\text{in.}^3/\text{rev}) \, N \, (\text{rev/min}) \qquad (6.3)$$

Since 1 gal $= 231$ in.^{3}, we may write

$$Q_T(\text{gpm}) = V_D N/231 \qquad (6.4)$$

Equation 6.4 shows that for a fixed displacement pump, the theoretical flow rate increases linearly with rotational speed as shown in Figure 6.3.

Since there must be clearance between the pump body and pressure plates and the rotating parts, leakage will take place. This causes a flow from the

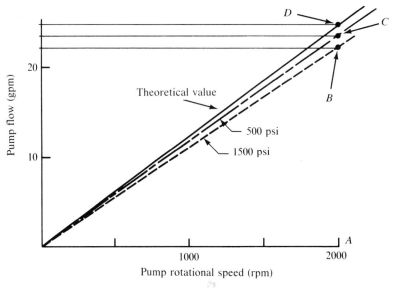

FIGURE 6.3
Pump rotational speed (rpm).

high-pressure port back to the inlet where the pressure is atmospheric. Therefore, the flow is less than theoretical, and the ratio of the actual to the theoretical flow is termed *volumetric efficiency,* defined as

$$e_V = \frac{Q_A}{Q_T} \times 100 \tag{6.5}$$

where $Q_A = Q_{Actual}$, and Q_T is given by Equation 6.4. Figure 6.3 shows for a typical pump the variation in actual flow at working pressures of 500 psi and 1500 psi in addition to the theoretical curve of pump flow and speed. You will see that the higher the pressure, the more so-called "slippage" flow and the lower the volumetric efficiency.

Example 6.2 ▶ A gear pump has external and internal diameters of gear teeth (width 1 in.) of 2.5 and 1.75 in. Calculate the volumetric displacement. If the delivery is actually 17.1 gpm at 1700 rpm, calculate the volumetric efficiency.

Solution:
From Equation 6.2,

$$V_D = \frac{\pi}{4}(2.5^2 - 1.75^2)1 = 2.503 \text{ in.}^3$$

From Equation 6.4, the theoretical flow rate is

$$Q_T = V_D\frac{N}{231} = \frac{(2.503)(1700)}{231}$$

$$Q_T = 18.4 \text{ gpm}$$

The volumetric efficiency is then, from Equation 6.5

$$e_V = \frac{17.1}{18.4} \times 100 = 93\% \qquad ◀$$

Example 6.3 ▶ Using Figure 6.3, compute the value of volumetric efficiency of the pump described there for delivery pressures of 500 psi and 1500 psi.

Solution:
The variation with rpm of theoretical delivery and delivery at 500 psi and 1500 psi are linear with a constant slope. Thus, for 500 psi

$$e_V = \frac{\text{slope of 500 psi line}}{\text{slope of theoretical line}} = \frac{AC}{AD}$$

$$e_V = 94.5\%$$

Similarly

$$e_V \text{ for 1500 psi} = \frac{AB}{AD}$$

$$e_V = 90\%$$ ◄

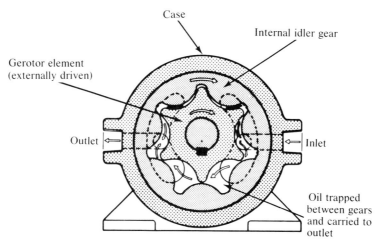

FIGURE 6.4
The gerotor type pump. (Courtesy Vickers, Incorporated)

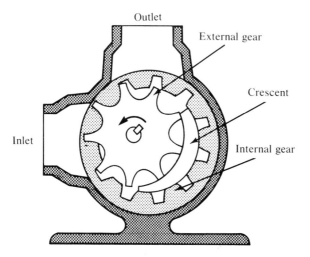

FIGURE 6.5
The crescent type pump. (Courtesy Vickers, Incorporated)

An internal gear pump may be the gerotor or the crescent type. In either case, the principle remains the same. The oil is conveyed between unmeshed gear teeth and is then forced by the remeshing of the teeth to exit the pump body, as shown in Figures 6.4 and 6.5.

Figure 6.6 is a photograph of a typical external gear pump. The illustration also shows the symbol used to represent a fixed displacement pump when drawing hydraulic circuits.

An example of a screw type pump is a so-called *three-screw configuration* (Figure 6.7). The central rotor is driven and turns in mesh with two idler rotors and the walls. Again, the inlet oil is brought into the pump because of the partial vacuum. It is then propelled axially along the walls of the casing and out the other end. Note here that the flow is free of pulsations and noise, and would be a good choice where the design demands such flows.

Vane pumps pump fluid by the action of vanes that slide in slots in a rotor, causing them to ride on the inner wall of the pump housing, called a cam ring. In the configuration shown in Figure 6.8, it is possible by means of a fixed eccentric shaft and an oval-shaped housing to provide force balance. If, however, variable displacement is required, a circular housing is usually employed with provision for changing the rotor eccentricity (Figure 6.9). The mode of

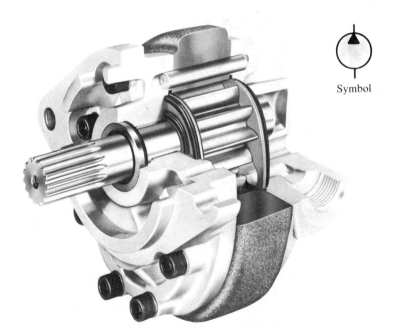

Symbol

FIGURE 6.6
A typical external gear pump, showing symbol used. (Courtesy Webster Electric Company, Inc.)

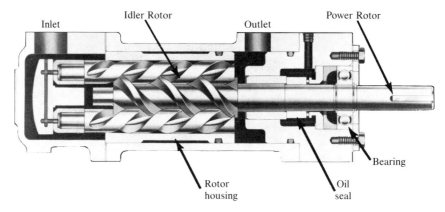

FIGURE 6.7
Axial flow three-screw pump. (Courtesy Transamerica Delaval Inc.)

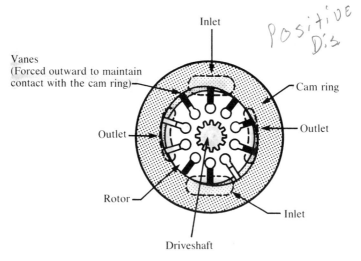

FIGURE 6.8
Vane pump balanced for force (fixed placement). (Courtesy Vickers, Incorporated)

action is to trap the inlet oil between the vanes and the outer case, delivering the oil at high pressure through the appropriate ports.

For the volumetric displacement, a more complex relationship exists than for the gear pump. However, as can be seen, the principle used in all hydraulic pumps is applied where the internal geometry is made to vary during its revolution. This change in geometry in turn causes oil to be drawn into the inlet and expelled at the outlet at the system working pressure. Volumetric efficiency usually can be obtained by measuring the volumetric displacement at very

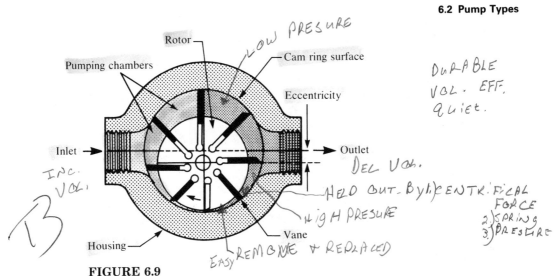

Handwritten annotations on figure: LOW PRESSURE; DURABLE VOL. EFF. QUIET.; INC. VOL; DEL VOL.; HELD OUT. BY 1) CENTRIFICAL FORCE 2) SPRING 3) PRESSURE; HIGH PRESSURE; EASY REMOVE & REPLACED; B

FIGURE 6.9
Variable delivery vane pump. (Courtesy Penton IPC Publications)

low delivery pressure and comparing this to values when high pressures are generated by the pump.

Example 6.4 ▶ A vane pump has its inlet connected to a reservoir of oil at atmospheric pressure, and the shaft is rotated by means of a hand crank. When the air had been expelled from the pump, 10 rotations of the shaft result in three pints of oil being delivered from the pump outlet. When the pump is rotated at 1500 rpm, the delivery is 50 gpm at the rated pressure. Determine the volumetric efficiency of the pump at this rated pressure.

Solution: First calculate the volumetric displacement. Since hand-cranking produces only a minute pressure, we can assume no leakage (or slip flow) is occurring. Thus

$$V_D = (3/8)(1/10) \frac{\text{gal}}{\text{rev}} = 0.0375 \frac{\text{gal}}{\text{rev}}$$

At 1500 rpm $\quad Q_T = (0.0375)(1500) = 56.25 \text{ gpm}$

Volumetric efficiency gives the ratio of actual to theoretical delivery. So

$$e_V = \frac{Q_A}{Q_T} \times 100 = \frac{50}{56.25} \times 100$$

$$e_V = 89\% \qquad\qquad ◀$$

Piston pumps are classified according to their configuration and may be of the radial or axial type. Both types convert rotary shaft motion to a reciprocating motion. In the radial piston pump a series of pistons are arranged radially in a cylinder block. The cylinder block rotates on a stationary pintle inside a

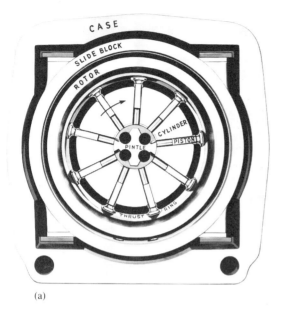

(a)

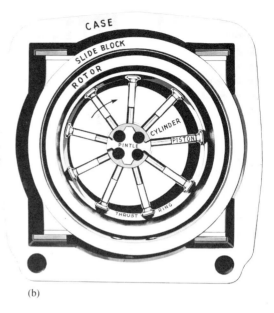

(b)

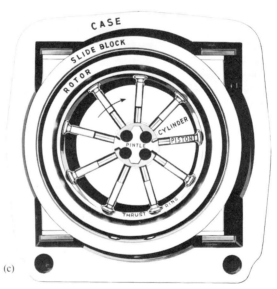

(c)

FIGURE 6.10

A radial piston pump. (Courtesy The Oilgear Company) (a) Slide block, rotor, and thrust ring assembly centerline is concentric with cylinder and pintle centerline. (Neutral: no fluid is delivered.) (b) Slide block, rotor, and thrust ring assembly centerline is moved to right of cylinder and pintle centerline. Fluid is delivered through lower pintle port. (c) Slide block, rotor, and thrust ring assembly centerline is moved to left of cylinder and pintle centerline. Fluid is delivered through upper pintle port.

circular thrust ring. As the block rotates, the pistons follow the inner surface of the ring, which is offset from the cylinder block centerline. By following this surface the pistons reciprocate in their bores. The oil being pumped is drawn into them as they move outwards, and discharged from them as they move inward. The diameter and number of pistons determine the pump displacement, which can be varied by moving the reaction ring. This changes its eccentricity and therefore its piston travel. The change in position may be done mechanically, hydraulically, or electrically. Flow reversal or zero-flow conditions may also be achieved. Figure 6.10 indicates how flow discharge and directions are changed.

There are two types of axial piston pumps: the swashplate pump and the bent-axis pump. In the swashplate design the cylinder block is rotated. Pistons in its bores are connected through slippers and a holding plate so that the slippers bear against an angled swashplate. As the block rotates, the pistons perform a reciprocating motion, the piston stroke being dependent upon the angle of the swashplate. In this pump, as in the case of the radial piston pump, the flow can be reversed. See Figure 6.11.

In the bent-axis pump the drive shaft rotates the pistons, which move against a stationary port plate. The axis of rotation of the barrel group makes an angle with the drive shaft. Thus, the distance between any one of the pistons and the port plate surface changes continuously during rotation: individual pistons move away from the port plate (sometimes called a *valving surface*) during one-half of the shaft revolution and towards it during the other half. The valving surface is ported so that oil inlet occurs while the pistons move away, and oil outlet occurs when the pistons move towards it. Although the

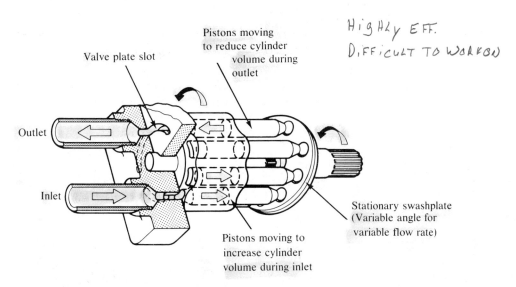

FIGURE 6.11
The swashplate axial piston pump. (Courtesy Vickers, Incorporated)

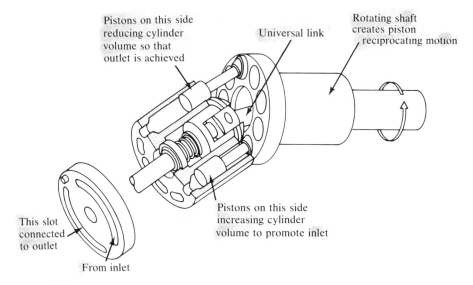

Pistons on this side reducing cylinder volume so that outlet is achieved

Universal link

Rotating shaft creates piston reciprocating motion

This slot connected to outlet

From inlet

Pistons on this side increasing cylinder volume to promote inlet

FIGURE 6.12
The bent-axis pump. (Courtesy Vickers, Incorporated)

flow quantity can be changed by changing the angle between the barrel group and the driveshaft, the flow cannot be reversed in this design (see Figure 6.12).

6.3 PUMP PERFORMANCE

In earlier sections we described how to express the operating properties of a pump in terms of its overall efficiency from Equation 6.1 and also its volumetric efficiency from Equation 6.5. The pump mechanical efficiency actually indicates the amount of input energy consumed to overcome losses caused by reasons other than leakage. These losses include friction in bearings and friction between other parts of the pump in relative motion. From an earlier description

$$e_M = \frac{\text{theoretical horsepower required to operate pump}}{\text{actual horsepower delivered to pump}} \times 100$$

or

$$e_M = \frac{\text{pump output horsepower (assuming no leakage)}}{\text{input horsepower delivered to pump}} \times 100 \quad \textbf{(6.6)}$$

The overall efficiency of the pump is given by

$$e_o = \frac{\text{pump output horsepower}}{\text{pump input horsepower}} \times 100$$

$$e_o = \frac{pQ_A}{T_A N} \times 3060 \times 100$$

From Equation 6.6

$$e_M = \frac{\text{pump output horsepower (assuming no leakage)}}{\text{pump input horsepower}} \times 100$$

and expressing e_M in terms of torque

$$e_M = \frac{T_T}{T_A} \times 100 = \frac{pQ_T}{T_A N} \times 3060 \times 100 \qquad \textbf{(6.6a)}$$

Also, we know that

$$e_V = \frac{Q_A}{Q_T} \times 100$$

But

$$e_o = \frac{pQ_A}{T_A N} \times 3060 \times 100 = \frac{p e_V Q_T}{T_A N (100)} \times 3060 \times 100$$

After substituting Equation 6.6a into Equation 6.1

$$e_o = \frac{e_V e_M}{100} \qquad \textbf{(6.7)}$$

We see that a simple relationship exists among the three pump efficiencies

overall efficiency = volumetric efficiency × mechanical efficiency

Example 6.5 ▶ A pump has a volumetric displacement of 3.5 in.[3] If it delivers 25 gpm at 1000 psi and the input required from the prime mover at a speed of 1740 rpm is 52 lb·ft, calculate (a) the overall pump efficiency and (b) the volumetric efficiency. Also, (c) calculate the theoretical torque required to operate the pump.

Solution: Use Equation 6.4 to derive the theoretical flow rate.

$$Q_T = \frac{V_D N}{231} = \frac{(3.5)(1740)}{231} = 26.4 \text{ gpm}$$

(b) Then solve for the volumetric efficiency.

$$e_V = \frac{Q_A}{Q_T} \times 100 = \frac{25.0}{26.4} \times 100 = 94.7\%$$

The mechanical efficiency is, from Equation 6.6a

$$e_M = \frac{p Q_T}{T_A N} \times 3060 \times 100$$

Remember, pressure p is in 1000s of psi. Then

$$e_M = \frac{1(26.4)}{(52)(1740)} \times 3060 \times 100 = 89.3\%$$

(a) Overall efficiency is, from Equation 6.7,

$$e_o = \frac{e_M e_V}{100} = \frac{94.7 \times 89.3}{100}$$

$$e_o = 84.6\%$$

(c) From Equation 6.6a

$$T_T = \frac{p Q_T (3060)}{N}$$

$$T_T = \frac{(1.0)(26.4)(3060)}{1740}$$

$$T_T = 46.40 \text{ lb} \cdot \text{ft} \qquad \blacktriangleleft$$

6.4 PERFORMANCE CURVES

The complete performance of a pump is usually placed on one graph. The curves show in a composite manner the effect that varying a basic variable (such as pump outlet pressure) has upon such quantities as efficiency. An example of such a composite graph is shown in Figure 6.13.

6.5 PUMP DRIVES

In the previous sections you saw that when oil is pumped through a pipe or hydraulic system and work performance is required of it, there is a resulting loss of fluid energy. How can we put the energy into a fluid so it can be utilized?

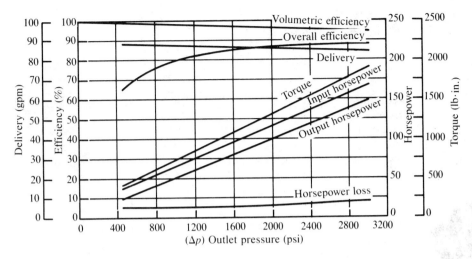

FIGURE 6.13
Typical performance curves of a pressure-compensated pump. (Courtesy Vickers, Incorporated)

For industrial hydraulic applications, the pumps are generally driven at synchronous speeds (1200 or 1800 rpm) by electric motors. Mobile hydraulic applications such as backhoes, tractors, and steering units are driven from engine take-off shafts for speeds varying between 1000 and 4000 rpm. For aerospace systems where geared-power take-offs from gas turbines and from high-speed electric motors are employed, speeds of up to 12,000 rpm are common. These high speeds in aircraft are necessary because weight considerations are critical. High power has to be generated from small pumps. In practice, friction and heat losses within the system are factors that limit the available power, forcing it to supply more energy than output.

6.6 DIFFERENT TASKS FOR DIFFERENT PUMP TYPES

The previous sections showed the different types of pumps that are manufactured and how they are supplied with energy. The performance of the various pump types are specified by the manufacturers in two ways. These are in tabular form and by use of illustrations, where volumetric efficiency, overall efficiency, and horsepower input are given as a function of delivery pressure and speed. A manufacturer invariably makes a number of different sizes so that work functions requiring differing capacities can be accomplished. Some types of pumps are better suited to perform a particular task than others. Table 6.1 compares different pump designs in terms of capacity, pressure, overall efficiency, weight, cost, and speed.

109

TABLE 6.1
Comparison of pumps.

Pump Type	Pressure Rating (psi)	Overall Efficiency (%)	Weight (lb/hp)	Cost ($/hp)	Rated Speed (rpm)
external gear	2000 to 3000	80 to 90	0.5	5 to 9	1200 to 2500
internal gear	500 to 2000	60 to 85	0.5	5 to 9	1200 to 2500
vane	1000 to 2000	80 to 95	0.5	7 to 35	1200 to 1800
axial piston	2000 to 10,000	90 to 98	0.25	8 to 60	1200 to 3600
radial piston	3000 to 10,000	85 to 95	0.35	7 to 40	1200 to 1800

Gear pumps are the least expensive, and are therefore used extensively for mobile equipment machine tools and unsophisticated aircraft applications. Their efficiency is reduced quickly by wear that occurs in the region of all the vital clearances. Improved performance can be obtained by use of helical type of gears.

Vane pumps are medium-priced and do not wear nearly as quickly as gear pumps, provided that clean hydraulic systems (to within manufacturer's specifications) are used in conjunction with them. Overall efficiency is sometimes at a disappointing level around 75 to 87 percent. Losses are due to leakage across the faces of the rotor and between the bronze wear plates and the pressure ring. Vane pumps can handle the relatively large flow rates but will whine at speeds greater than approximately 2400 rpm.

Piston pumps are the most expensive of all. They are also most efficient due to the ability to provide close-fitting pistons and components. They are used when especially high performance is required. In experimental aircraft circuits, pressures of up to 10,000 psi have been used. Another benefit of these pumps (especially in aircraft use) is their long life expectancy.

Example 6.6 ▶ A gear pump is selected to drive a load at a rate of 2 in./s using a jack of diameter 4.5 in. Compute the range of powers you would be expected to supply to the input shaft of the pump.

Solution:

$$\text{flow rate} = \underbrace{\frac{\pi}{4}(4.5)^2(\text{in.}^2)}_{\text{jack area}} \times \underbrace{2(\text{in./s})}_{\text{speed}} = 31.8 \text{ in.}^3/\text{s}$$

In terms of gpm

$$Q = 31.8 \, \frac{\text{in.}^3}{\text{s}} \times \frac{1 \text{ gal}}{231 \text{ in.}^3} \times \frac{60\text{s}}{1 \text{ min}} = 8.26 \text{ gpm}$$

range of pressure (from Table 6.1) = 2000 to 3000 psi

$$\text{minimum power of the range} = \frac{8.26 \times 2}{1.714} = 9.64 \text{ hp}$$

$$\text{maximum power of the range} = \frac{8.26 \times 3}{1.714} = 14.5 \text{ hp}$$

For minimum power of the range, use an overall efficiency of 90%.

$$\text{minimum power} = \frac{9.64}{0.9} = 10.7 \text{ hp}$$

For maximum power calculations, Table 6.1 indicates use of a minimum efficiency of 80%; therefore

$$\text{maximum power} = \frac{14.5}{0.8} = 18.1 \text{ hp}$$

Thus, the expected power requirement for the task assumed varies between 10.7 hp and 18.1 hp. ◄

6.7 CAVITATION IN PUMPS

The phenomenon of cavitation limits to some extent the height to which oil can be raised to enter the pump inlet port. It actually occurs when the pressure of the oil at some point reaches that of the saturated vapor, which for hydraulic oils varies with temperature and viscosity. However, a representative value could be a vacuum of around 12 psi. In other words, when the oil pressure is reduced from the atmospheric pressure by 12 psi, local boiling will occur.

When the oil boils, bubbles of oil vapor and entrained air come out of solution and are carried through the pump. Increased pressure causes the bubbles to be reduced in size. If they do collapse completely near a surface, considerable damage is done to that surface. The pump also becomes very noisy.

One reason for reduced local oil pressure is that the pump attempts to suck oil through a vertical distance to its inlet. However, as was shown in Chapter 3, reduction of pressure is caused by increased velocity or by a reduction in the ambient static oil pressure, such as that achieved by increasing the height of the oil in the earth's atmosphere. In the design of hydraulic circuits, cavitation is prevented or reduced by the following procedures.

1. Remove air bubbles from the inlet oil, because they help form vapor bubbles.

2. Choose pumps with inlets designed to have small entry velocities.
3. Avoid low-diameter inlet pipes (and thus high-inlet velocities) to a pump.
4. Charge the inlet oil (i.e., increase its pressure with an auxiliary pump) so that higher inlet velocities may be used without reducing the oil pressure to the vapor pressure as a result.

Example 6.7 ▶ The vapor pressure of the oil to be used in a certain pump with a flow rate of 10 gpm is 3.5 psia. The atmospheric pressure is 14.7 psi, and an additional 3 psi is given to the inlet oil by means of a charge pump. Assuming the inlet to the pump is designed for low passage velocity and is 10 ft vertically above the reservoir surface, calculate the minimum pipe diameter if the specific gravity, sg, for the oil is 0.85.

Solution: In this instance, it is easier to work in absolute units of the slug, foot, and second. We assume no frictional head loss, so the Bernoulli equation is used between the reservoir surface and inlet. For this equation (assuming the velocity in the oil reservoir is zero)

$$\frac{p_r}{\gamma_o} + \frac{p_c}{\gamma_o} = \frac{p_i}{\gamma_o} + \frac{v^2}{2g} + h_{p-R}$$

where
p_r = reservoir pressure = 14.7 psi

v = velocity in pipe (required)

p_c = charge pressure = 3 psi

h_{p-R} = required lift for oil = 10 ft

p_i = maximum value of inlet pressure permissible = 3.5 psi

$\gamma_o = \gamma_{oil} = 62.4 \times 0.85 = 53.04$ lb/ft^3

Substituting, therefore

$$\frac{(14.7 + 3.0)144}{53.04} = \frac{3.5(144)}{53.04} + \frac{v^2}{2(32.2)} + 10$$

$$v^2 = 2(32.2)\left[\frac{(14.7 + 3 - 3.5)144}{53.04} - 10\right]$$

$$v^2 = 1838 \text{ ft}^2/\text{s}^2$$

$$v = 42.9 \text{ ft/s}$$

This shows that the maximum flow speed possible before cavitation occurs is 42.9 ft/s. (Actually, due to such factors as lack of uniform velocities, the maximum permissible velocity will be somewhat less than this.) However, assuming that 42.9 ft/s is barely possible and that the flow rate is 10 gpm, then by use of Equation 3.2, the diameter of the pipe is given by

$$D^2 = \frac{0.408 \times 10}{42.9} = 0.095 \text{ in.}^2$$

$$D = 0.31 \text{ in.}$$

A pipe of 0.31 in. diameter would be the absolute minimum that could be used to assure the inlet pressure of 3.5 psi would always be exceeded. Now let's use the knowledge acquired in this and previous chapters to perform calculations on a comprehensive problem. ◀

Example 6.8 ▶ Figure 6.14 shows part of a hydraulic system in which oil of specific gravity 0.83 is pumped along a 1 in. diameter (pipe 1) to a cylinder where it operates a piston ram. The piston is to move at 3 in./s against a steady force of 600 lb. The piston area is 30 in.², and the rod area is 6 in.² Pressure gauges are fitted in A in pipe 1 and at B in the 1 in. diameter, pipe 2; the length of each pipe from gauge to piston is 40 ft. Calculate the loss of pressure along each pipe if the viscosity of the oil is 2.0 poise. Determine the reading of the gauge at A if the reading of the gauge at B is 4 psi.

Solution: We first find the rate of flow from the load velocity requirement. Thus, on pump side (pipe 1)

oil displacement per second = 3(in./s)(30)in.² = 90 in.³/s

Reynolds number in pipe 1 and pipe 2 must first be calculated. As an illustration we calculate the Reynolds number based upon the cgs system and then use this number to calculate pressures in the British system. This procedure will work only because Reynolds number is dimensionless. As an exercise you should work out Reynolds number using the British system. (You should get the same answer for Reynolds number and for your final answer.)

$$v_1 = 3[\text{in./s}] [30(4)/\pi(1)^2] = 114.6 \text{ in./s} = 291.1 \text{ cm/s}$$

$$v_2 = 3[\text{in./s}] [(30 - 6) (4)/\pi(1)^2] = 91.7 \text{ in./s} = 232.8 \text{ cm/s}$$

$$\rho = 0.83 (1.0) [\text{gm/cm}^3]$$

$$\mu = 2[\text{gm/cm·s}]$$

$$d = 2.54 \text{ cm}$$

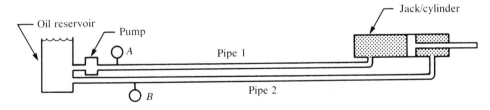

FIGURE 6.14
Diagram of hydraulic system.

$$\text{Re}_1 = \frac{(0.83)\,(291.1)\,(2.54)}{2} = 306.8 \qquad \text{Re}_2 = 244.6$$

Head loss between A and p_1 (where p_1 is pressure in the left-hand side of the jack/cylinder) from Equation 3.16 is

$$H_A = \frac{64}{306.8} \times \frac{40(12)}{1}\left[\frac{114.6}{12}\right]^2 \times \frac{1}{2(32.2)} = 141.8 \text{ ft of oil}$$

$$\text{pressure loss from } A \text{ to } p_1 = \frac{(0.83)\,(62.4)\,(141.8)}{144} = 51 \text{ psi}$$

Similarly

$$\text{pressure loss from } p_2 \text{ to } B = \frac{(0.83)\,(62.4)\,(141.8)}{144}\left(\frac{91.7}{114.6}\right)^2\left(\frac{306.8}{244.6}\right) = 41.0 \text{ psi}$$

From the load requirement

$$p_1 \times 30 - p_2 \times 24 = 600 \text{ lb}$$

$$p_B = 4 \text{ psi}$$

$$\therefore p_2 = 45 \text{ psi}$$

$$p_1 = \frac{600 + (24 \times 45)}{30} = 56 \text{ psi}$$

Therefore, reading at gauge $A = 56.0 + 51.0 = 107.0$ psi ◀

6.8 PUMP SELECTION

Design necessitates continual checking to ensure that the task specifications are being met. Therefore, many different pump circuit component combinations are used so the job of design and the required task can be performed in an efficient manner best meeting the specifications at a minimum cost. The following are some helpful pointers for selecting the most suitable pump for a stated work requirement.

1. The first consideration is that of the actuator requirements. At this stage the detailed geometry of the task is defined (see Chapter 8). From these considerations, a concept of flow volumes necessary from the pump is obtained. Since there will usually be a time limit involved for the task performance, the flow rate requirements can be calculated.
2. The required loads determine the choice of system pressure. Values of pressure and of flow rate will generally determine which pump type is

required (see Table 6.1). They also enable us to make initial evaluations of power requirements.

3. The type of prime mover available will determine the speed and type of pump to be selected. For instance, is a gear, vane, or piston pump needed?

4. Next, a circuit is drawn to determine the type of expected losses. As a result of these calculations, together with system pressure and flow rate, the total power required can be decided. In addition, the type of pump to be used determines how much power is required from the prime mover, since we should have a good idea as to the pump's overall efficiency.

5. Detailed design of the circuit (described in Chapter 9) can now take place, and a cost analysis can be conducted.

6. As will be discussed in Chapter 9, several cycles of iterative design procedure should be followed to optimize the design.

T3

Summary

Different types of positive displacement pumps are described in this chapter. Their mode of action and their operating efficiencies are defined and described, and ways in which pump data may be presented are outlined. Methods of driving pumps are given, together with an idea as to their application tasks. The relative costs of the different pumps are shown so that decisions in the design of hydraulic circuits can be made. Finally, the phenomenon of cavitation is described and means of eliminating it or reducing its effects are given.

Questions and Problems

1. What is the purpose of the pump in a hydraulic system?
2. Describe how inlet conditions of a pump are accomplished. List the three main types of pump and draw diagrams to illustrate your explanation.
3. Compute the overall efficiency of a pump if it requires 5.5 hp to drive it when it is delivering 3.9 gpm at a pressure of 2100 psi.
4. A pump will deliver 9 gpm at 1300 rpm. If its volumetric efficiency is 95%, calculate the volumetric displacement.
5. A pump with an overall efficiency of 82% rotates at 1200 rpm, delivers oil at a pressure of 2500 psi, and requires a torque of 82 lb·ft on the shaft to drive it. Determine the flow rate from the pump in gpm and in in.^{3}/s.
6. Two identical gears are to be used in a gear pump to provide a volumetric displacement of 3.5 in.^{3} If the outer and inner diameters of the gear teeth are 2.75 in. and 2.00 in., respectively, calculate the width of gear tooth required.

115

7. It was found that the pump in Problem 6 gave an actual delivery of 24.3 gpm when rotating at 1800 rpm. Calculate the volumetric efficiency. If the mechanical efficiency is equal to 92%, calculate the torque required to rotate the pump if the delivery pressure is 2250 psi.

8. Calculate the mechanical efficiency of a pump with a volumetric displacement of 0.42 in.^{3}/rev when it is working at 1200 psi and rotating at 1100 rpm. You are given that 1.65 hp are required to drive the pump under these conditions, and the volumetric efficiency is 96%.

9. A pump operates at 2500 psi and delivers 7 gpm. If the overall efficiency of the pump is 85%, calculate the power required to drive it.

10. A vane pump delivers 11.45 gpm at 1850 rpm if it is operated at a pressure of 1500 psi. If its volumetric displacement is 1.50 in.^{3}/rev, calculate the volumetric efficiency.

11. The hydraulic jack shown in the following diagram is operated by pumping oil of viscosity 1 poise around the closed circuit of ¼ in. diameter pipe. The lengths of the supply and return lines are both 60 ft. The area of the cylinder is 6 in.^{2} and of the piston rod 2 in.^{2} Compute the pressure difference between the inlet and delivery sides of the pump when the piston operates at a rate of 1 in./s against a steady load of 4 tons. Then, find the pump output horsepower and determine what percentage of the work done is lost due to viscous friction in the pipelines. (Specific gravity of oil is 0.85.)

Load = 4 tons

Pump

Closed circuit

Suggested Learning Activities

Obtain a gear pump and connect the inlet to a reservoir of oil. Attach a cranked handle to the pump drive. When oil that is free of air is delivered, measure the amount of oil delivered for a given number of revolutions (a suitable number would be 10). From the results calculate the volumetric displacement. Dismantle the pump and measure the inner and outer diameter of the gear wheels so the theoretical volumetric displacement can be calculated from Equation 6.2. Compare your two values.

Suggested Readings

Esposito, Anthony. *Fluid Power with Applications*. Englewood Cliffs, N.J.: Prentice-Hall, 1980.

Sullivan, James A. *Fluid Power: Theory and Applications*, 2nd ed. Reston, Va.: Reston, 1982.

Wolansky, William; Nagohosian, John; and Henke, Russell W. *Fundamentals of Fluid Power*. Boston: Houghton Mifflin, 1977.

CHAPTER 7

Energy Modulation Devices

Introduction

Control is the most important design consideration in any fluid power system. If the control components are not properly selected, installed, and operated, the entire system's efficiency, safety, and reliability will suffer. The two control factors that primarily affect energy transfer are fluid pressure and fluid flow. Whereas pressure control valves affect the potential energy level of the fluid in the hydraulic system, the flow control valves regulate the quantity of fluid flowing past a reference point per unit of time. Thus, the product of fluid pressure and fluid flow rate is the power transferred by the circuit fluid for a given time interval.

There are three basic types of control devices: pressure control valves, flow control valves, and directional control valves. This chapter describes these three types of control valves and illustrates their construction, operation, and variations of application. Pressure control valves protect the system against excessive pressure due to sudden surges or gradual build-up of pressure. Flow control valves regulate fluid flow rate in various lines of a hydraulic circuit. A variable pump can also be used to control fluid flow rate. Directional control valves determine the path through which the fluid flows within a given circuit.

Valve: A device that controls fluid flow, direction, pressure, or flow rate.

Valve, Directional Control: A valve whose primary function is to direct or prevent flow through selected passages.

Valve, Flow Control: A valve whose primary function is to control flow rate.

Valve, Pressure Control: A valve whose primary function is to control pressure.

7.1 PRESSURE CONTROL VALVES

Pressure level in a hydraulic power transmission system provides several key control functions:

1. Maintains pressure levels necessary to accomplish desired work.
2. Limits the maximum output force of the hydraulic system.
3. Maintains efficiency of pressure ranges throughout the various branch circuits of a hydraulic system.
4. Maintains pressure level required for specific functions in a given time.

There are two modes of pressure control used in fluid power circuits:

1. Regulate direct control of the pressure level. This can be achieved with relief valves to control maximum pressure; with reducing valves to control pressure at some level below the maximum system pressure; and with the pressure-compensated, variable-displacement pump.
2. Secondary control of the pressure level. This isolates the secondary circuit from the primary circuit until the set pressure is reached. Such controls include sequence valves to switch flow to a secondary circuit when the fluid pressure in the primary circuit has reached a preset pressure level. Unloading valves are also included in this category. Their purpose is to bypass pump and/or accumulator flow to the reservoir after a certain pressure level in the system downstream of a check valve is reached.

7.2 DIRECT-ACTING RELIEF VALVE

The most widely used type of pressure control valve in a hydraulic system is the pressure relief valve. It is a normally-closed valve whose function is to limit the pressure to a specified maximum pressure level. When pressure exceeds the maximum value, the pressure relief valve allows the fluid to flow back to the reservoir. In effect, the relief valve first limits the maximum system pressure, which protects the pump and other system components such as piping, hoses, tubing, and actuators. The relief valve also limits the maximum output force of the hydraulic system. This valve is adjustable within given ranges and may be set and locked to operate at a required pressure level (Figure 7.1).

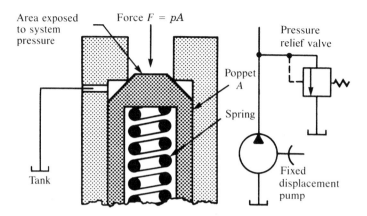

FIGURE 7.1
Direct-acting relief valve is the most commonly used means of controlling maximum operating system pressure.

The construction and actuation of a direct-acting relief valve are relatively simple. An adjustable mechanical spring holds a poppet, plug, ball, or sliding spool closed against an orifice that prevents flow from the pump to the reservoir. When the fluid pressure levels acting on the valve element overcome the bias force of the mechanical spring, the element, the poppet, ball, or sliding spool will crack open and allow fluid to return to the reservoir.

The pressure valve design is such that the sliding spool, poppet, or ball is held in closed position by an adjustable mechanical spring. Opposite this spring force is an element area that is exposed to the system pressure. Compression springs exert a force proportional to the amount they are compressed. A numerical value is assigned to compression springs based upon their stiffness. Depending on the stiffness of the spring and the extent of compression, the increase in force can be very high per inch of compression. Since the spring constant value is expressed in units of lb/in. and the compression length is also expressed in inches, the spring force can be determined by the equation

$$\text{spring force} = kX = \text{lb}$$

where $k = \text{spring constant (lb/in.)}$
$X = \text{compression length (in.)}$

The fluid pressure acting on the element surface area will create a balancing force until it exceeds the spring force (preset maximum) pressure level to maintain an open condition.

$$\text{force} = \text{pressure} \frac{(\text{lb})}{(\text{in.})^2} \times \text{area (in.)}^2 = \text{lb}$$

See Figure 7.2. A direct-acting relief valve operates over a pressure range, rather than at one specific pressure. At some pressure level, referred to as the *cracking pressure,* the valve opens slightly to allow a small portion of the total pump flow to bypass to the reservoir. At this state, it reduces the flow to the circuit by this small amount of fluid. As the pressure level continues to increase, the valve element opens more and allows larger amounts to bypass, until the pressure level reaches the predetermined pressure setting and opens the valve sufficiently to bypass all pump output back to the reservoir. The spring reseats the element (ball, poppet, or sliding spool) when enough fluid is

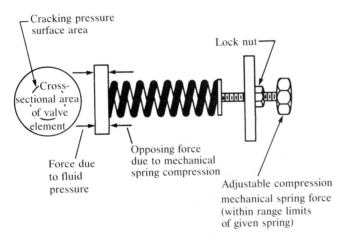

FIGURE 7.2
Basic balancing of forces between the spring and the fluid pressure.

bypassed to drop the system pressure below the setting of the valve spring force. Thus, the relief valve protects circuit components in the system from excessive and damaging pressure effects.

Notice that while the direct-acting relief valve operates within a preset pressure range, it can be adjusted to open at any pressure within the valve design range. Aside from this adjustment feature, the valve operates like the fixed valve once it is set and locked into position.

7.3 THE PILOT-OPERATED RELIEF VALVE

The pilot-operated relief valve performs the same function as the direct-acting valve. It protects the system components from excessive pressure and limits the maximum output force. Pilot-operated relief valves increase pressure sensitivity by reducing the pressure override that is usually encountered with valves that use only the direct-acting force of system pressure against the valve.

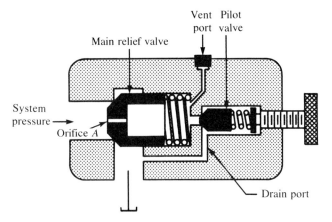

FIGURE 7.3
Operation of a pilot-operated relief valve.

For applications requiring valves capable of relieving large flows (75 gpm or more) with small pressure differences, pilot-operated relief valves are preferred. This valve operates in two stages. The pilot stage consists of a small, spring-biased relief valve built into the body of the main relief valve, which acts as a trigger to control the main relief valve (Figure 7.4). The main relief valve is normally closed when the inlet oil pressure is below the setting of the main valve spring force. Orifice A in the main valve poppet (Figure 7.3) permits system fluid pressure to act on a larger area on the spring side of the poppet. The sum of these forces (main spring bias and system pressure) keeps the poppet seated and prevents fluid flow to the reservoir. At the same time, the pilot valve is also closed. Pressure in passage B is the same as system pressure, and is less than the preset pressure valve of the pilot valve spring.

As the system pressure rises, the pressure in passage B also rises. When it reaches the setting of a pilot valve, the pilot valve poppet opens. Oil is released behind the main valve poppet through passage B and the drain port. The resulting pressure drop across orifice A in the main valve poppet causes it to open and allows oil to bypass to the reservoir, preventing any rise in inlet system pressure.

The valve will close again when inlet fluid pressure drops below the valve setting. It is the pilot valve poppet that controls the pressure differential in passage B, which controls the main valve to remain closed or open. Pilot-operated valves do not respond until the system attains the upper operational system pressure; thus, the efficiency of the system is enhanced because less oil is released to the reservoir. This type of relief valve is better suited to high-pressure and high-volume applications. These valves also maintain a more constant pressure while relieving the pump directly to the reservoir due to the sensitivity to pressure differentials.

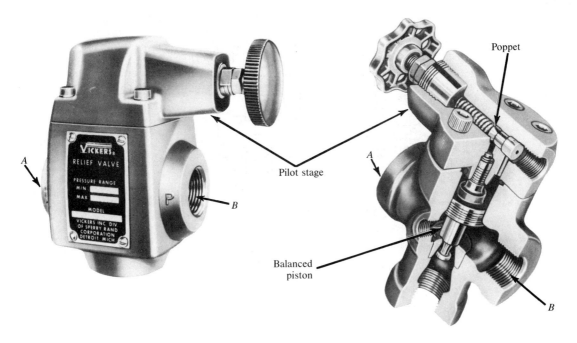

FIGURE 7.4
External and cutaway views of a compound relief valve. (Courtesy Vickers, Incorporated)

The compound valve illustrated in Figure 7.5 includes a spool element in the main valve and a poppet in the pilot valve. Figure 7.5a indicates how fluid pressure acts on both sides of piston (1), which is normally held closed on its seat by a relatively light bias spring (2). This is made possible by the presence of an orifice (C) in the land of the spool. Figure 7.5b shows that when the pressure increases sufficiently to move the pilot poppet valve (4) from its seat, fluid behind the main spool is able to drain. The resulting pressure imbalance on the main spool (1) causes it to move in a direction towards the lower pressure. The result is that piston spring (2) is compressed and lets the main spool rise, allowing the fluid to return to the reservoir and preventing a further rise in pressure. In this type valve, the pressure setting is adjusted with the adjustment screw that increases or decreases the fluid pressure required to unseat the pilot poppet valve.

While relief valves protect the circuit components, they also offer resistance to flow, particularly high-pressure and high-volume flows. This causes considerable generation of heat and loss of energy transfer efficiency. Directional valves, having larger spools and passages, discharge fluids to the reservoir more efficiently, particularly when the intervals between cycles of actuators are long.

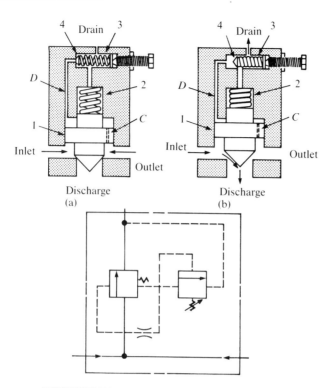

FIGURE 7.5
Compound pilot drained pressure relief valve. (Courtesy Vickers, Incorporated)

7.4 PRESSURE-REDUCING VALVES

Pressure-reducing valves directly supply branch circuits with fluid at a lower pressure than the system pressure value. By reducing pressure in the secondary circuit, it is possible to independently limit the output force to that necessary in the primary circuit. Pressure-reducing valves are normally-open, two-way valves, which to close receive their actuating signals from downstream fluid pressure (or secondary circuit). See Figure 7.6. To ensure proper functioning, pressure-reducing valves should be drained back to the tank. This practice prevents downstream pressure buildup.

The direct-acting, pressure-reducing valve shown in Figure 7.7 operates as follows: Fluid flows unobstructed from inlet to outlet in Figure 7.7a, since this is a normally-open valve. The pressure-reducing spool is held open by spring while any leakage past the spool will return to the drain. As the downstream pressure at the outlet increases, it acts through passage *(E)* against the end of the spool. Since the other end of the spool is open to drain, the unbalanced spool will compress the spring as shown in Figure 7.7b. As pressure increases, the spring will compress more, further restricting the flow from the

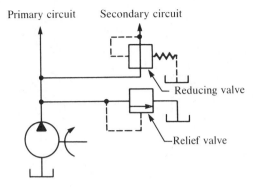

FIGURE 7.6
Pressure-reducing valve supplies a secondary circuit fluid pressure below that of a primary circuit.

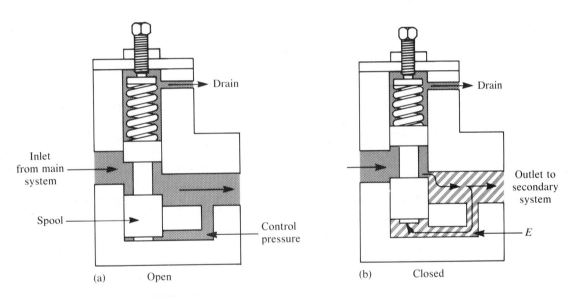

FIGURE 7.7
Pressure-reducing valve.

inlet to the outlet port. As the pressure drops at the outlet port and at passage *(E)*, the spool will be returned by the mechanical force of the spring to its open position. The pressure at the outlet is maintained at the preset pressure level independent of the inlet pressure, and at some value less than the primary circuit fluid pressure. The pressurized fluid at the reducing valve's outlet has less capacity for doing work than it had at its inlet. The loss in pressure at the downstream side is a loss in force potential, which reflects higher heat losses. If the pressure differences are too large between the primary and secondary cir-

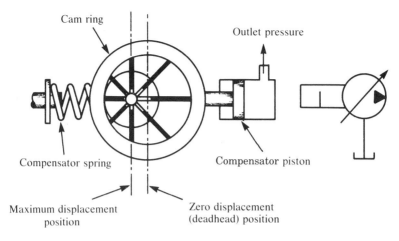

FIGURE 7.8
Schematic of pressure-compensated, variable displacement vane pump.

cuits (particularly in high-flow system applications), it may be advisable to use a second pump.

Another method of directly controlling pressure in a circuit is to use a pressure-compensated pump. This type of pump is capable of delivering a pressure span specified by the pressure requirements of the system. These pumps have a pressure-versus-volumetric displacement characteristic. However, different pressures can be supplied by the pressure-sensing capability of such a pump. As illustrated in Figure 7.8, the vanes of this pump are driven at a fixed rotor position and are contained by an adjustable cam ring. As shown in Figure 7.8, an increase in system pressure to reach the cutoff pressure will exert a fluid pressure on the compensator piston. This moves the cam ring to the left and reduces the eccentricity, thereby reducing the displacement of oil.

As the pressure continues to rise, the cam ring will ultimately reach the full neutral position, where the rotor and vanes continue to rotate with no displacement of oil to the system. The pressure adjustment is made manually by adjusting the spring compression.

A pressure-compensated pump can function as a system relief valve. Because no oil is being throttled across a pressure valve, less energy is lost through heat generation in this type of pump application. The primary circuit pressure is controlled by the preset pressure level called the cutoff pressure.

7.5 SECONDARY CONTROL PRESSURE VALVES

This category of pressure control valve generally controls pressure in secondary circuits at a somewhat lower level than the primary circuit pressure.

Examples include sequence valves and unloading valves. A sequence valve is a control device used to switch flow to a secondary circuit when fluid pressure in the primary circuit has reached the preset pressure level (Figure 7.9). Sequence valves are used in circuits with two or more actuators, which requires moving these actuators in a definite sequence.

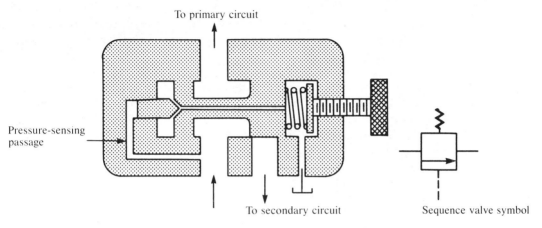

FIGURE 7.9
Major parts of a sequence valve.

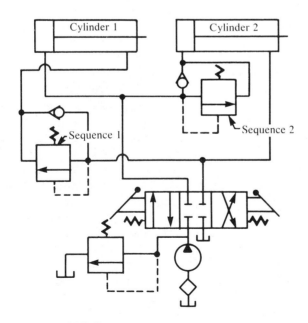

FIGURE 7.10
Sequence circuit.

The pressure-sensing sequence valve's primary function is to direct flow in a predetermined sequence. This valve operates on the principle that when the main system pressure overcomes the spring setting, the valve spool moves upward, allowing oil flow from the primary to the secondary port.

The spool is normally held in a closed position by an adjustable spring. As the system pressure reaches the sequence valve pressure setting, the signal from the direct (or an external pilot) source causes the spool to shift upward and open flow from the secondary port. Fluid is then directed to the next actuator. When the circuit requirements are such that reverse flow through a sequence valve is necessary, an internal check valve allows this to occur.

When a pressure control valve operates and allows oil to flow through the secondary port to perform work, it is important that an external drain is provided so the oil can be bled from the spring chamber of the valve. Otherwise, the accumulated, incompressible fluid in the spring chamber will prevent the spool from shifting. Trace the fluid flow patterns in Figure 7.10 for the extension and retraction cycles of cylinders 1 and 2. This circuit, for example, could operate the clamping and unclamping of a piece of stock in a drill vise, and feeding the spindle and retracting it on a drill press.

7.6 UNLOADING VALVE

Unloading valves are normally-closed, two-way valves externally piloted and often internally drained (Figure 7.11).

Figure 7.12 illustrates a schematic of an unloading valve used to unload the pump connected to port A when the pressure is maintained at the predetermined value at port X. The high-flow poppet is controlled by the spring-loaded ball and the pressure applied to port X. Flow at port A is blocked by the poppet at low pressure. The pressure signal at A passes through the orifice in the main poppet to the topside area and on to the pilot ball. There is no flow through these sections of the valve until the pressure exceeds the maximum permitted by the adjustably-set, spring-loaded ball. When the pilot ball is unseated, the pressure behind the main poppet is reduced. The main poppet lifts, causing the flow to pass from port A (pump) to port B (reservoir). The system pressure signal to port X acts against the unloading piston, forcing the ball further off the seat. This causes the topside pressure on the main poppet to reach its very low value. This opens it far enough to allow greater flow from A to B with a very low pressure drop as long as signal pressure at X is maintained. The pilot ball will reseat and the main poppet will close when the pressure at port X falls to approximately 90 percent of the maximum pressure setting of the spring-loaded ball.

This valve allows an unloading circuit to reach an adjustable pressure setting and then discharge the pump flow to the reservoir at minimum resis-

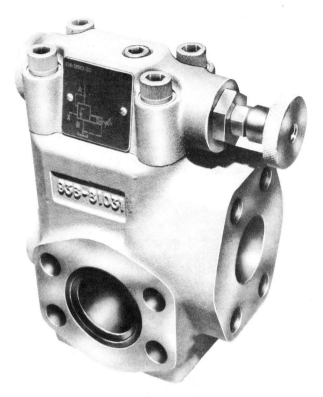

FIGURE 7.11
Unloading valve: externally piloted, internally drained. (Courtesy Abex
Corporation)

tance for as long as pilot pressure is maintained on the valve from a remote
source (Figure 7.12). For example, you can compute the horsepower loss across
a pressure relief valve with a setting of 1000 psi in a hydraulic circuit if all the
oil returns to the tank over the relief valve from a 10 gpm pump.

$$HP = \frac{\text{pressure} \times \text{flow}}{1714}$$

$$HP = \frac{pQ}{1714} = \frac{1000 \times 10}{1714} = 5.85$$

Yet, if an unloading valve is used to unload the same capacity pump, the
hydraulic horsepower consumed would be considerably less if the pump discharge pressure (during unloading cycle) equals 50 psi.

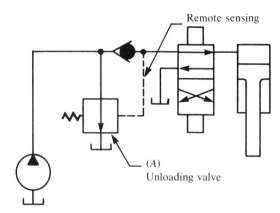

Remote sensing

(A)
Unloading valve

FIGURE 7.12
Unloading valve (A) opens when cylinder bottoms at the end of the stroke.

$$HP = \frac{50 \times 10}{1714} = 0.2917$$

Not only is less power drawn from the prime mover, but fluid flowing over the unloading valve generates proportionally less heat during the unloading cycle.

When fluid flow passes through the unloading valve, the pump has essentially no load. Therefore, it is consuming and developing a minimum of horsepower. This condition minimizes heat generation during the unloading valve bypass of fluid flow. Shifting the directional control valve to reverse the direction of the piston rod (Figure 7.12) causes a pressure drop in the circuit, allowing the unloading valve to reset.

The unloading valve function differs from the pressure relief valve. With a pressure relief valve in a circuit that serves as a primary pressure control, the pump is delivering full pump flow at the pressure relief valve's predetermined setting. Therefore, the pump is operating at higher horsepower conditions with significant heat losses.

The efficiency of an unloading circuit depends on the minimum pressure at which the high-volume pump circulates oil. The high-low circuits provide two output forces by using two different pressure capacity pumps (Figure 7.13). Such a system can deliver low force or torque and high actuator speed because of the high flow. Then, on a remote pressure signal, it can shift to a high force or torque and slow actuator speed by unloading the high-flow, low-pressure pump A. This type of application is frequently used on a rapid approach, slow-squeeze cycle such as a stamping press.

The unloading valve will unload the high-flow, low-pressure pump as soon as the system reaches a preset pressure as it encounters the squeeze force of the press. Prior to this pressure setting, both pumps supply fluid to the circuit. After the high-flow, low-pressure pump *(A)* is unloaded, only the high-

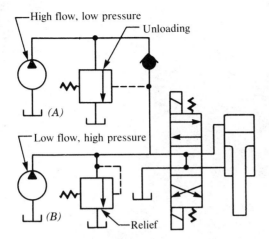

FIGURE 7.13
High-low circuit. Each of two fixed displacement pumps is protected by a separate relief valve.

pressure, low-flow pump *(B)* supplies the flow to exert a large force and a slow feed to the press cycle.

Another common application of the unloading valve is found in an accumulator circuit (Figure 7.14). The unloading valve with an external pilot is able to accept a pressure signal from an accumulator, thereby diverting pump flow to the reservoir when the accumulator is fully charged.

As the accumulator discharges somewhere between 10 and 25 percent below full-charge pressure, the unloading valve will close, causing the pump fluid flow to pass through a check valve *(A)* to recharge the accumulator and/or to supply fluid to the circuit along with the flow from the accumulator.

The unloading valve provides three functions in an accumulator circuit: It

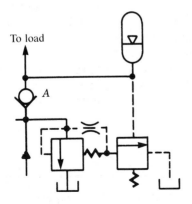

FIGURE 7.14
Typical accumulator loading circuit.

131

FIGURE 7.15
Accumulator unloading valve with solenoid venting. (Courtesy The Rexroth
Corporation)

limits maximum system pressure; unloads the pump to reservoir when the
accumulator reaches the desired pressure; and reloads the pump to bring the
accumulator up to full charge after a predetermined minimum pressure in the
circuit is reached and it remotely senses system pressure. The function of an
unloading valve in an accumulator should not be confused with an accumulator
safety valve. After an accumulator is charged (Figure 7.14), the unloading
valve has no means of bleeding the accumulator charge to the reservoir if the
system is turned off. As a precaution, accumulators need to be discharged prior
to any maintenance on components in the accumulator circuit.

7.7 FLOW CONTROL VALVE

A flow control valve meters fluid flow (Figure 7.17). A simple example of this is
a needle valve. A complex example is a pressure and temperature-compensated
valve that controls the rate of flow independent of system fluid pressure and

temperature. The deceleration valve gradually reduces flow to slow down a hydraulic motor or cylinder. A flow divider directs the flow from a single source into two or more branch circuits by controlling the fluid flow rate to regulate the speed of hydraulic cylinders or motors. The needle valve illustrated in Figure 7.16 allows the fine adjustment of the quantity of flow through the orifice by increasing or decreasing the orifice opening by adjusting the needle valve against the matched seat.

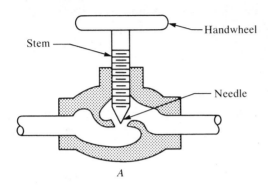

FIGURE 7.16
Flow control needle valve.

FIGURE 7.17
Adjustable flow control valve. (Courtesy Teledyne Republic)

Meter-in Circuit

Flow control circuits effect speed control of actuators. In some applications it may be essential to control the rate of the extension stroke of a cylinder. For example, a hydraulic table feed on a surface grinder would require controlled rate of travel.

In the meter-in circuit illustrated in Figure 7.18, the flow control between the directional valve and the cap end of the cylinder controls the rate of fluid flow into the cylinder, and therefore the output velocity of the piston. This circuit is used with resistance loads.

Meter-out Circuit

Metering fluid from the actuator (Figure 7.19) provides a restrictive action that can result in positive control of such machine tools as drill bits, taps, or milling cutters. These devices tend to advance by their own cutting action, which causes damage either to the tool or the work. It is this damage that the circuit is designed to eliminate.

A metering-out circuit prevents external forces from affecting the actuator's rate of movement. In a metering-out circuit, the flow control valve provides the necessary control to resist overrunning loads. In a meter-out speed control circuit, the pump operates against a relatively constant pressure, consisting of the combined load reaction and the flow control valve back-pressure.

Bleed-off

In this type of flow control circuit, a portion of the pump delivery is bypassed to the tank at system pressure. This circuit, with a constant flow pump, may

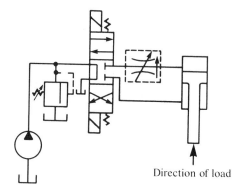

Direction of load

FIGURE 7.18
Meter-in flow control circuit.

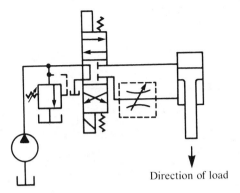

FIGURE 7.19
Meter-out flow control circuit.

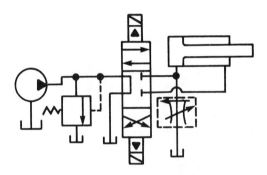

FIGURE 7.20
Bleed-off flow control circuit.

provide more output than required to maintain the desired output speed of the actuator. By diverting some portion of the flow to the tank, the desired speed of the actuator can be maintained.

In the bypass, diversion, or priority function, the circuit pressure level is established by the load on the primary actuator, rather than excess fluid being passed over a relief valve at a maximum set pressure. In Figure 7.20 note that this circuit provides speed adjustment of the actuator around some average value. It does not allow speed control over the entire range, nor does it provide positive load control.

Regenerative Circuit

The regenerative circuit (Figure 7.21) includes a special configuration directional control valve spool to allow both ends of the cylinder to be connected in parallel while one port (A) of the directional valve is blocked. The design of the

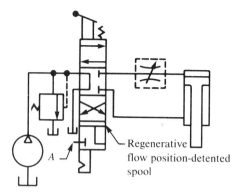

FIGURE 7.21
Regenerative circuit with regenerative position directional control valve.

circuit is intended to recycle the fluid by bypassing the reservoir and flowing directly to the blind or cap end of the cylinder. Because the cap end area is larger than the rod end, the forces with the same pressure will be greater on the cap end. This causes the cylinder piston to extend against the load reaction.

When the spool is shifted to the detented position, the speed for the extension stroke is faster than for a regular double-acting cylinder. This is because flow from the rod end (Q_R) regenerates with the pump flow (Q_P) to create a total flow rate (Q_T), which is the combined flow rate available to the cap end. Therefore, $Q_T = Q_R + Q_P$. If, for example, the ratio of the area of the cap end to that of the rod end is 2:1, the piston velocity will be the same in both directions.

Flow Divider Valves

Flow divider valves are designed to receive one input flow and split it into two or more output flows. The valve can deliver equal flow rates in each stream or a predetermined ratio of flow rates (Figure 7.22). Flow divider devices are intended to synchronize actuators as illustrated in Figure 7.22.

There are essentially two types of flow dividers. The first is an adjustable orifice divider. The second type is the rotary flow divider, which consists of several gear fluid motors connected in parallel (Figure 7.23). This type of flow divider is designed to divide one input flow *(Q)* into proportional, multi-branch output flows (Q_1, Q_2, Q_3, and Q_4).

Since all motor elements are mounted on the same shaft and turn at the same speed, the output stream flow rates are proportional to the displacement capacity of each motor. The total displacement is $Q_T = Q_1 + Q_2 + Q_3 + Q_4$. Depending on the number and displacement ratio of each gear motor, the flow rate outputs will total the input flow if slippage is not present.

The circuit designer must not allow pressure differentials to vary too much in individual branch circuits. In this way, the designed proportional

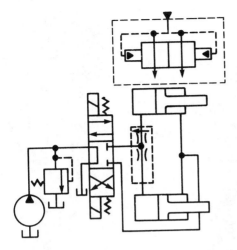

FIGURE 7.22
Flow divider splits input flow into multiple output flows.

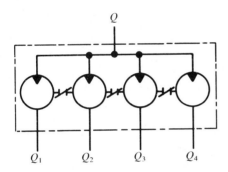

FIGURE 7.23
Rotary gear motor flow divider mounted in parallel on a common shaft.

flows can be maintained because slippage is minimized by sustained pressure differential tolerances. One limitation of the gear motor flow divider is that the control can only be varied with the speed of the shaft, as the gear displacement per revolution remains fixed for each motor mounted on the common shaft (Figure 7.24).

It is important to remember that flow control valves and devices control the rate of movement (velocity) of an actuator. Controlling fluid flow allows the designer to alter the rates of reciprocating or rotary speeds of actuators, accelerate or decelerate loads, or control the speeds of different actuators in a multiple actuator circuit. Actuators can also be synchronized.

Other variables that influence and affect actuator speed accuracy include pressure, temperature, and actuator efficiency. For example, the speed of an actuator will drop as the load pressure increases. Unless a pressure compen-

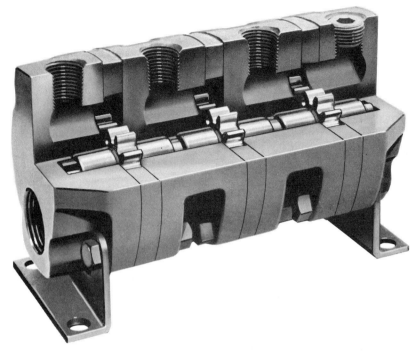

FIGURE 7.24
Functional gear motor flow divider. (Courtesy Delta Power Hydraulic
Company)

sated-control valve is used, this condition will continue. Temperature also has
an effect on fluid viscosity. The colder the oil, the more resistance to flow is
encountered across the control valve orifice. Actuator efficiency is also related
to volumetric efficiency or slippage. The higher the pressure drop across the
clearance fits in the actuator, the greater the reduction in speed due to fluid
slippage.

7.8 DIRECTIONAL CONTROL VALVES

Directional valves have a primary function to direct or prevent fluid flow
through selected circuit passages during a particular time. These types of control valves may vary in design from a simple check valve to a complex multiple
stacked valve package used to operate a number of circuits from one hydraulic
source. Most of the complex directional valves can be operated to assume any of
an infinite number of positions between the valve's minimum and maximum
limits. Therefore, it is possible to divert fluid flow and to also control flow.

Directional control valves are classified as one-way, two-way, three-way, and four-way, depending upon the number of combinations of passages to conduct the fluid. For example, a check valve is considered a one-way valve because the fluid can only flow through one path of the valve body (Figure 7.25). Fluid under pressure is free to flow only against the poppet by unseating it against the mechanical force of the spring. This directional valve allows flow in one direction but prevents flow in the opposite direction.

Two-way valves (or on-off valves) are used primarily as shutoff valves. Such valves may be constructed with a seating action utilizing a disc, a ball, a gate, or some other element to regulate the flow path.

The rotary-plate valves illustrated in Figure 7.26 allow an operator to direct flow by aligning the fluid passage. Flow may be prevented by moving the handle until the disc ports are blocked.

A three-way directional valve pressurizes and exhausts one port. Essentially it includes the supply port, the exhaust port, and the outlet or working port that directs flow to an actuator or to another valve (Figure 7.27). The three-way directional valve has three ports. In position 1, supply is directed to port A. In position 2, port A is open to exhaust.

A four-way directional valve pressurizes and exhausts two ports (Figure 7.28). It contains a supply port, an exhaust port, and two outlet ports. Observe that the fluid can be directed to flow through the valve in four ways, depending upon the position of the spool. Spool position 1 illustrates flow from P to port A and port B returning to T. While position 2 shifts to flow from P to port B, port A returns to T.

The two-way valve can be used to start and stop flow. A three-way valve is used to operate an actuator in one direction only. The return of the actuator is achieved by other means such as the weight of an aggregate carrier. The hydraulic fluid used for the lifting stroke is directed to exhaust to the reservoir by the operation of a three-way directional valve.

A four-way valve is used for bidirectional travel. An example is controlling the actuator on a front-loader bucket during the lowering or raising

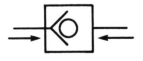

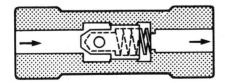

FIGURE 7.25
Operation of a check valve.

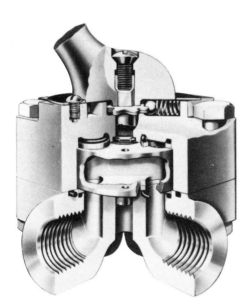

FIGURE 7.26
Rotary-plate valve. (Courtesy Teledyne Republic)

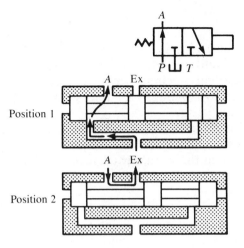

FIGURE 7.27
Three-way sliding spool directional valve.

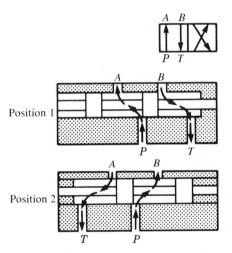

FIGURE 7.28
Four-way directional spool valve.

motions. Positive control is maintained by a four-way valve. The choice of the number of ways of a directional valve will depend on the specific routing of the fluid to a circuit or branch circuits. There are many possible configurations, as well as combining a number of valves to achieve specific routing patterns.

141

Position

Directional valves are described according to their position. The position determines the number of flow conditions the valve can maintain. Symbolically, each envelope of a directional valve contains the flow pattern in a given position. Positions can be altered by moving a ball, rotary disc, or sliding spool or plate to cover or uncover fluid ports.

When the valve spool is in flow position 1, the fluid flows from the pump through port A to the blind end of the cylinder. Return flow from the rod end is directed through port B back to the reservoir (Figure 7.29). While the valve spool is held in this position against the spring, the fluid flow will cause the piston to extend from the cylinder.

As the solenoid is de-energized the mechanical spring will shift the spool to its normal position, causing a flow reversal and thereby retracting the piston. With this two-position, four-way directional valve, it is possible only to extend or retract a cylinder or alternately control the direction of a fluid motor. As soon as the piston is fully extended or retracted, the pump output will be bypassed over the relief valve until the directional valve spool is shifted to a new position.

Three-position, four-way directional valves contain a variety of flow configurations in the center position. An open-center directional valve cannot hold clamping pressure, as the fluid is allowed to bypass over the open center back to the reservoir. In a front-loader cylinder application circuit using a closed-center valve, once the bucket is raised and the directional control valve is left in

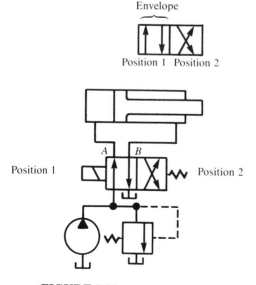

FIGURE 7.29
Graphic symbol represents a two-position, four-way directional valve.

the center position, it will remain raised, as all ports are blocked. A tandem-center directional valve performs the same functions in a front loader circuit. However, the fluid will bypass over the directional control spool to the reservoir instead of the relief valve as it would with a closed center. A wide variety of special configurations are manufactured for specific circuit applications (Figure 7.30). Some sliding spool valves have interchangeable spool capabilities to alter the possible flow pattern.

Actuation

Directional control valves may be actuated by a variety of methods (Figure 7.31). Actuation is the method of moving the valve element from one position to another.

Early directional control valves were actuated manually: physically moving a handle, pushing a button, or stepping on a foot pedal. Most of these early valves had spools that were spring-returned or spring-centered. As directional control valve designs improved to allow more complex and accurate control, the mechanical control yielded to electric, fluid, and electronic actuators, including programmable controllers. Frequently some combinations of these actuators were used to provide the needed accuracy and speed of control response. A variety of increasingly sophisticated methods of actuation have been developed (Figure 7.32). The range of actuation of directional valves may vary from

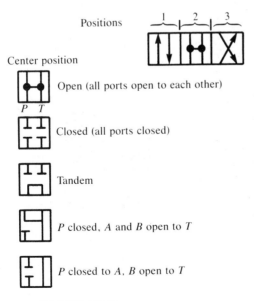

FIGURE 7.30
Center flow configurations of three-position directional valve.

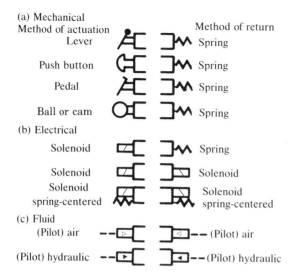

FIGURE 7.31
Methods of actuators of directional control valves.

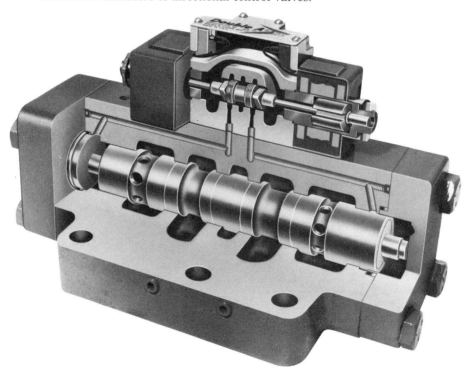

FIGURE 7.32
Solenoid-actuated, spring-centered directional control valve. (Courtesy Double A Products Company, Manchester, Michigan)

simple manual controls to solenoid, relays, pilot-operated to programmable controllers.

Servo-valves are used to accomplish more critical and precise control in such applications as automation, space vehicles, and other advanced hydraulic systems. A servo system is a combination of elements used to accomplish control of a source of power. The output of the system is fed back for comparison with the input and the difference between these quantities is used in regulating the power transfer (Figure 7.33).

A servo-valve is designed to control hydraulic output more accurately by regulating velocity, acceleration, or position in response to an electric input signal. The amount and the direction of fluid flow are related to the polarity and magnitude of the electric signal. Servo-valves are used in closed-loop systems where the controller is continuously responsive to the cylinder or fluid motor that is regulated by feedback.

An input electric signal to the servo-valve must be converted into mechanical motion to actuate the control function. In Figure 7.34, the flow of fluid to or from the load actuator is proportional to the input current signal of the force motor. Programmable controllers require programming changes and may require additional modules. However, the designer has much greater flexibility with input and output devices. The output devices typically include relay coils, valve solenoids, motor starter coils, and others. The central processing unit (CPU) programs instructions from memory and feedback signals tell-

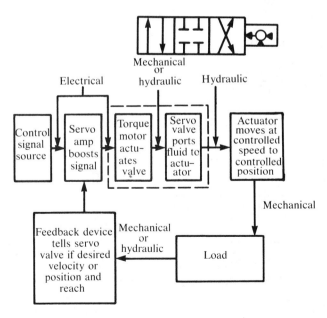

FIGURE 7.33
Block diagram of an electro-hydraulic servo-valve system.

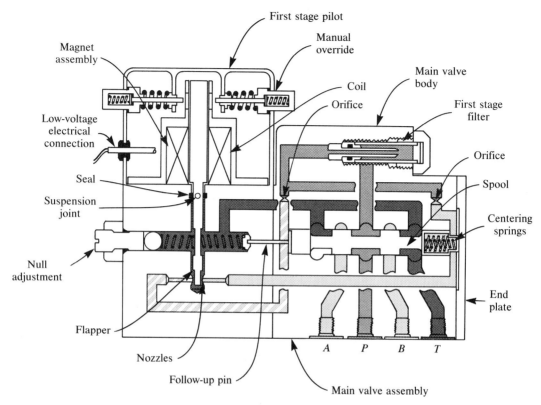

FIGURE 7.34
Servo-control valve. (Courtesy Commercial Shearing)

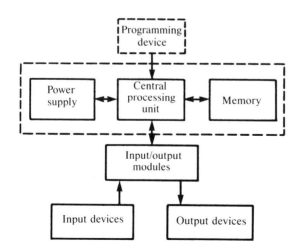

FIGURE 7.35
Block diagram of microprocessor programmable controller.

ing the status of the input/output devices that control the operating sequence of the output devices (Figure 7.35). Programmable controllers will be discussed in Chapter 10 in greater detail.

Directional valves have become complex devices. It is perhaps in the area of controls that the interfacing of mechanical, electrical, and fluid will yield greater accuracy, speed, and position of actuators. Programmable controllers have significantly contributed to the flexibility of inherent control functions of hydraulic circuits.

The selection of hydraulic valves is very much a function of the variables to be controlled. Essentially, control valves are designed to regulate pressure, flow, and direction. Some valves are entirely mechanical devices that can regulate pressure, flow, or direction by the design features of a valve. Other valves may use remote pressure sensing to create a switching action of flow through the valve.

Valves also vary in pressure, flow, and directional capabilities. To design efficient, reliable, and safe circuits, it is critical to determine loads, cycle profiles, select pressure and flow requirements, determine force/torque profiles, and calculate power profiles. Efficiency, speed, safety, and cost are also important considerations in determining which control valves are essential.

Manufacturer's specifications will be helpful in selecting appropriate control valves. It is also possible to design and fabricate simple circuits to accomplish certain actuator outputs or very complex, more accurate, and safer circuits to perform a similar sequence of controls. The overall important considerations in selecting control valves are safety, efficient use of energy, positive control, system reliability, and justifiable cost.

Summary

Hydraulic valves are devices designed to control pressure, flow rate, or the direction of the fluid in a circuit.

The function of a pressure control valve is to control pressure within desirable limits in a circuit or branch of a circuit. The pressure-relief control valve's primary function is to provide pressure limitation in a circuit, thereby protecting components against excessive system pressure.

The sequence valve has the primary function of directing flow in a predetermined sequence in reaction to pressure differences sensed from the load actuator. The unloading valve is designed to permit a constant volume pump to operate at a minimum load when fluid is bypassed over the unloading valve.

Directional valves may direct, divert, combine, stop, or reverse flows in a circuit. The flow divider may divide a single output into several equal or proportional outputs to various branches of a circuit.

Remote sensing, programmable controllers, and increased precision of speed and position of actuators are being achieved with the more complex control valves. This is perhaps the area of greatest promise for expanded use of hydraulic systems.

Questions and Problems

Hm. Wk.

1. Describe the main purpose of a pressure relief valve.
2. How does a compound relief valve differ from a simple relief valve?
3. What is the primary function of a sequence valve?
4. What is the primary function of an unloading valve?
5. What is the purpose of a pressure-reducing valve?
6. Compute the horsepower loss across a relief valve if it is set for 1000 psi and returns all flow back to the reservoir from a 20 gpm fixed displacement pump.
7. Compute the horsepower loss across an unloading valve if it is set for 50 psi and returns all flow back to the reservoir from a 20 gpm fixed displacement pump.
8. Distinguish between ways (3-way or 4-way) and positions in directional control valves.
9. What is the function of a flow divider?
10. Sketch and label the ANSI symbol for the following methods of actuation of a directional valve: manual, solenoid, hydraulic pilot, and servo.

Suggested Learning Activities

1. Locate the relief valve on the fluid power trainer in your laboratory. Determine if it is a simple or compound relief valve.
2. Draw a schematic circuit for a clamp and drill operation. If provisions are made on the laboratory trainer, fabricate and test the circuit.
3. Connect a flow control valve into the trainer circuit and observe the velocity change of a piston as you adjust the flow control valve.
4. Disassemble a three-position, four-way, spring-centered, solenoid actuated directional valve and observe the flow paths. Draw an ANSI symbol for this valve.
5. Read an article on programmable controllers and write a one-page summary of how this system works.

Suggested Readings

Esposito, Anthony. *Fluid Power with Applications.* Englewood Cliffs, N.J.: 1980, pp. 219–51.

Frankenfield, Tom. *Using Industrial Hydraulics,* 2nd ed. Cleveland, Oh.: Hydraulics and Pneumatics Magazine, 1979, chapters 2–7.

Henke, Russell W. "Fluid Power Systems and Circuits". *Hydraulics and Pneumatics Magazine.* Cleveland, Oh., 1983, pp. 71–99.

Pippenger, John J. *Hydraulic Valves and Controls: Selection and Application.* New York: Dekker, 1984.

Sullivan, James A. *Fluid Power: Theory and Applications,* 2nd ed. Reston, Va.: Reston, 1982.

Wolansky, William; Nagohosian, John; and Henke, Russell W. *Fundamentals of Fluid Power.* Boston: Houghton Mifflin, 1977, pp. 141–70.

Yeaple, Frank. *Fluid Power Design Handbook.* New York: Dekker, 1984, pp. 137–56.

CHAPTER 8

Fluid Actuators

Introduction

This chapter includes information about energy output devices, namely cylinders and motors. Hydraulic actuators are of two basic categories— linear or rotary. A cylinder is a linear actuator; a motor is a rotary actuator. An actuator is a device that converts fluid energy into a mechanical force or motion. By the end of this chapter you will become familiar with the principles of operation, construction, and proper application of hydraulic actuators, and learn how they may be used in circuit applications, which will be presented in Chapter 9.

Key Terms

Actuator: A device that converts fluid energy into a mechanical force or motion.

Cap: A cylinder-end closure that completely covers the bore area.

Cushion: A device that provides controlled resistance to motion.

Cylinder: A device that converts fluid energy into linear mechanical force or motion. It usually consists of a movable element such as a piston, plunger, or a ram operating within a cylindrical bore.

Double-acting Cylinder: A device in which fluid force can be applied to the movable element in either direction.

Head: A cylinder-end closure that covers the differential area between the bore area and rod area (sometimes referred to as the *rod end*).

Motor: A device that converts fluid energy into mechanical force and motion. It usually provides rotary output mechanical motion, which may be continuous rotation or limited rotation.

Single-acting Cylinder: A device in which fluid force can be applied to the movable element in only one direction.

Stroke: The distance through which the load must be moved determines the stroke. In turn, this factor will influence cylinder selection.

8.1 LINEAR HYDRAULIC ACTUATORS

A hydraulic cylinder converts fluid energy into mechanical force or power to perform useful work. The principle of actuator operation is based upon the fact that liquids are incompressible for most practical purposes (less than 0.05% at 1000 psi). This incompressibility of liquids permits the piston to transfer power within an efficient hydraulic system (Figure 8.1).

The simplest type of linear actuator is the single-acting, or ram form of cylinder. The single-acting cylinder can exert a force only during the extension stroke (Figure 8.2). As fluid enters the cap (blind) port, it forces the piston to extend. In a dump truck hoist, for example, the single-acting cylinder raises the loaded aggregate box. The weight of the box causes the piston to retract. Some single-acting cylinders have a mechanical spring to retract the piston when the pump is vented to the reservoir. They do not retract by a hydraulic force. Note that a single-acting cylinder has only one port. Hydraulic fluid enters and returns through the same port.

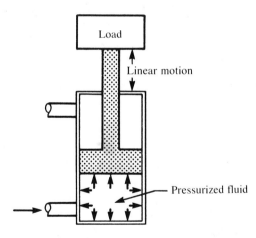

FIGURE 8.1
Linear actuator.

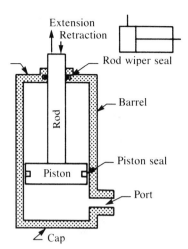

FIGURE 8.2
Single-acting cylinder.

8.2 DOUBLE-ACTING CYLINDER

A double-acting cylinder has two ports, permitting fluid under pressure to extend or retract the piston (Figure 8.3). This cylinder is referred to as a *differential cylinder,* because the areas on which the fluid acts are greater on the

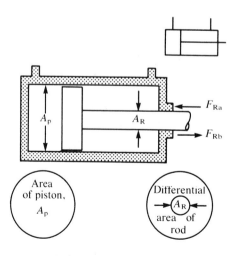

FIGURE 8.3
Double-acting cylinder.

cap end (extension stroke) than on the head end (retraction stroke). Using F_{Ra} as the force of the reaction load for the extension stroke and F_{Rb} as the force of the reaction load on the retraction stroke, the system pressure for both strokes can be determined as follows. For the extension stroke the equation is

$$\text{system pressure } (p) = \frac{\text{extension load reaction } (F_{Rb})}{A_P}$$

For the retraction stroke the equation is

$$\text{system pressure } (p) = \frac{\text{retraction load reaction } (F_{Ra})}{A_P - A_R}$$

Example 8.1 ▶ Suppose $A_P = 7.0$ in.^{2} and $A_R = 1$ in.^{2}, respectively. Assume the load reactions equal 700 lb in both directions. Compute the pressure for (a) the extension stroke and (b) the retraction stroke.

Solution:

$$\text{(a)} \quad p = \frac{700}{7} = 100 \text{ psi}$$

$$\text{(b)} \quad p = \frac{700}{7 - 1} = 116.66 \text{ psi} \quad ◀$$

Greater pressure is required to move the same force (load) on the retraction stroke, because the effective area that the pressurized fluid acts upon is reduced by whatever the cross-sectional area of the rod is for the given cylinder.

The velocity for the extension and retraction strokes would also be different for a given displacement pump. Recall from Chapter 3 that the velocity of a cylinder as it extends or retracts depends on how much fluid (gallons per minute) is delivered to the cylinder for a given diameter or bore of a cylinder. The velocity varies directly with the gallons per minute (gpm) of liquid delivery, and inversely with the net area of the cylinder (Figure 8.3). Mathematically, these relationships are expressed as

$$v \text{ (ft/min)} = \frac{(\text{gal/min}) \times 19.25}{A \text{ (in.}^2)}$$

The factor of 19.25 is derived by converting the distance that one gallon of oil (231 in.^{3}) will move a piston of given area from inches to feet per minute. That is, $231/12 = 19.25$.

Example 8.2 ▶   If a 5 gpm pump extends and retracts the double-acting cylinder illustrated in Figure 8.3, it will be determined that the velocity also varies for (a) the extension stroke and (b) the retraction stroke.

Solution: (a)

$$v = \frac{\text{gpm} \times 19.25}{A}$$

$$v = \frac{5 \times 19.25}{7}$$

$$v = 13.75 \text{ ft/min}$$

(b)

$$v = \frac{5 \times 19.25}{6}$$

$$v = 16 \text{ ft/min}$$

A double-acting cylinder will therefore retract faster than it extends for a given displacement pump. ◀

    For double-acting cylinders the differential area ratios are determined by the size of rod used for a given bore-size cylinder. The ratio of the bore size and rod size depends on the construction, application, and length of stroke. The material used and the stop tubing also influence this ratio. The double-end rod cylinder has a rod protruding from both ends (Figure 8.4). If both rods are the same diameter, the cylinder provides equal force and velocity in both directions. It is also possible to have rods of different sizes for special applications of the double-end rod cylinder.

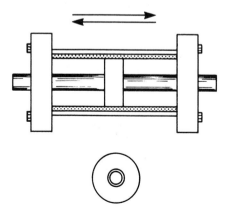

FIGURE 8.4
Double-acting, double-end rod cylinder.

8.3 CONSTRUCTION

Although there is a great variety of hydraulic cylinders manufactured for numerous applications, a cylinder is basically a barrel constructed in the form of a tubular housing that has been honed to a fine finish on the inside (Figure 8.5). The cylinder has a closely fitted piston, some cup packing to seal against leakage between the piston and the barrel, ports located in the end caps, tie-rods or some thread form of fastening, wipers and rod seals, and a cushion. Rods

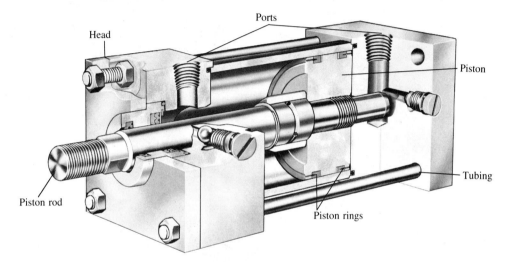

FIGURE 8.5
Cross section of an industrial hydraulic cylinder. (Courtesy The S-P Manufacturing Corporation)

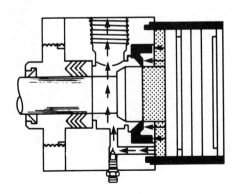

FIGURE 8.6
Adjustable metered bypass cushion flow.

155

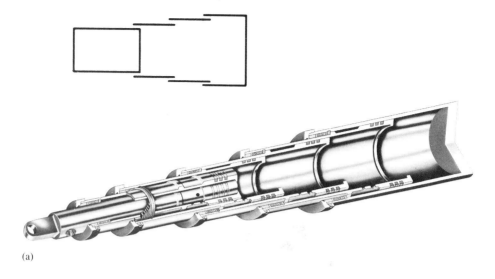

(a)

(b)

FIGURE 8.7
(a) Telescopic cylinder. (Courtesy Babcock & Wilcox) (b) Use of telescopic
cylinder for heavy road construction work. (Courtesy Babcock & Wilcox)

may be hard-chrome-plated to resist pitting and scoring, and to prolong seal and wiper life. Seals and packing materials are very important to prevent external leakage.

Cushioning hydraulic cylinders helps reduce the inertia forces and decelerates the piston near the end of the stroke. Heavily loaded or rapidly moving pistons must be cushioned to protect the piston against impact damage. Several methods can be used to hydraulically decelerate the piston at the end of the stroke (Figure 8.6). One method is to use a tapered plunger on the rod. Upon entering a counterbore opening in the cylinder cap end, the tapered plunger closes the normal discharge port, forcing the oil to leave through a small orifice in the end cap. A small adjustable needle valve increases or decreases the orifice opening, and hence the flow rate, which controls the deceleration rate of the piston.

Telescopic Cylinder

This cylinder type consists of a nested group of hollow sections with a piston that telescope together in the retracted position (Figure 8.7a). Telescopic cylinders are used on installations where the working stroke is long, but the retracted length must be very short. Typical applications include hydraulic elevators, dump truck hoists, and scissor cargo hoists (Figure 8.7b). One drawback of a telescopic cylinder design is that the area of each successive inner cylinder is less than the preceding one, thus proportionately changing the force exerted and speed, if they are not compensated for. Telescopic cylinders can be single-acting or double-acting.

8.4 BODY STYLE

Cylinders are also manufactured in different body styles. There are four major body styles, each designed for specific applications. The oldest and most common body style is the tie-rod design (Figure 8.8).

Four or more tie rods hold the end caps against the tubular housing. High tensile steel rods are used to absorb the internal energy stresses. Heavy duty cylinders with this body style are used in machine tool industries, mobile equipment, and automotive applications.

The threaded-construction cylinder has the end caps threaded to fit the internal thread of the tubular housing (Figure 8.8). This is a relatively new body style made possible by improved materials and machining capabilities. Such cylinders are used in aircraft, food processing, and transfer line applications. The mill type cylinder has the end caps bolted to mated flanges on the cylinder body (Figure 8.8). Recessed socket-head threaded bolts may be used to

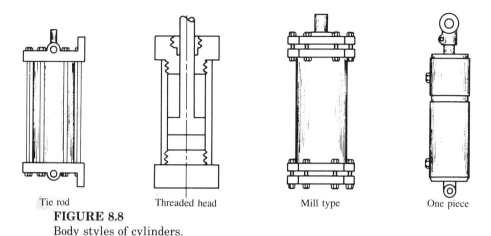

Tie rod   Threaded head   Mill type   One piece

FIGURE 8.8
Body styles of cylinders.

fasten the caps to the flanges. Because the flanges are frequently welded to the barrel, these cylinders are usually thick-walled and widely used in the steel and rubber manufacturing industries.

The one-piece cylinder may be a misnomer. Frequently the cap end and the head end are welded to the body. Such cylinders are used more extensively in mobile equipment and farm equipment. These cylinders can be mass produced at lower costs and are discarded when they develop extensive internal leakage.

8.5 MOUNTING CONFIGURATIONS

The types of mountings on cylinders are numerous, and they can accommodate a wide variety of applications (Figure 8.9). The most common mountings are lugs, flange, trunnion, clevis, and extended tie rods. One of the important considerations in selecting a particular mounting style is whether the major force applied to the load will result in tension or compression of the piston rod (Figure 8.10). The rod length-to-diameter ratio should not exceed about a 6:1 proportion at full extension. This prevents the rod from buckling due to compression or tensile shock forces.

Alignment of the rod with the resistive load is another important consideration in selecting cylinder mounts. Misalignment or nonaxial loading also tend to place unnecessary loads on the rod and rod guides, bushings, or bearings. Because of the immense loads and extreme forces induced by the rod, there are large stresses on the rod at full extension. Center lugs, center trunnions, and clevis arrangements tend to help keep the rod or shaft in line with the load.

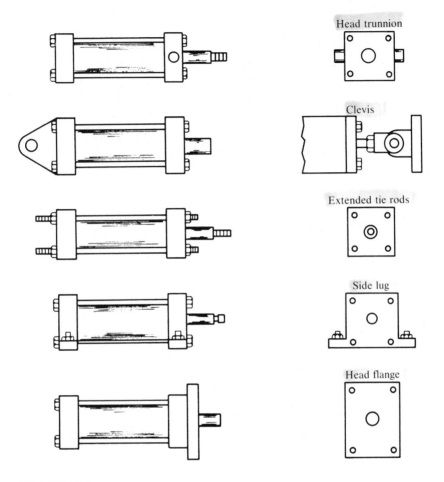

FIGURE 8.9
Mounts for square-head, tie-rod cylinders.

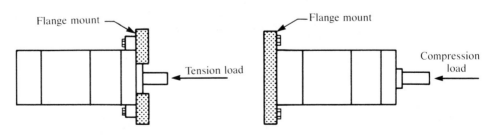

FIGURE 8.10
Tension and compression loads.

159

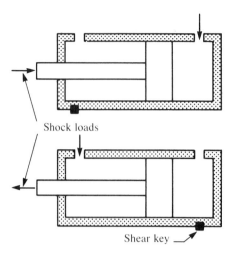

FIGURE 8.11
Shear keys to absorb shock.

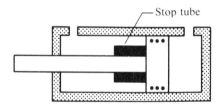

FIGURE 8.12
Application of stop tubing.

The placement of a shear key to absorb shock should exploit the maximum cylinder elasticity. A key should be used on one end only, as illustrated in Figure 8.11. A key takes up shear loads, provides accurate alignment of the cylinders, and simplifies installation and servicing.

For cylinders with long strokes, a stop tube can be used to regulate the length of stroke (Figure 8.12). More importantly, the stop tube provides support to reduce the bending moment of the rod when the rod is fully extended.

8.6 MECHANICAL LINKAGES

A cylinder can be attached to a load with various mechanical linkages as illustrated in Figure 8.13. The appropriate mechanical lineage can transform linear

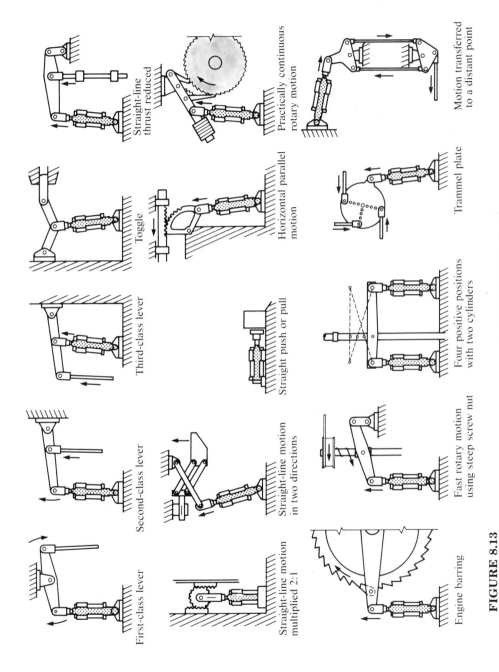

Straight-line
thrust reduced

Practically continuous
rotary motion

Motion transferred
to a distant point

Toggle

Horizontal parallel
motion

Trammel plate

Third-class lever

Straight push or pull

Four positive positions
with two cylinders

Second-class lever

Straight-line motion
in two directions

Fast rotary motion
using steep screw nut

First-class lever

Straight-line motion
multiplied 2:1

Engine barring

FIGURE 8.13
Application of cylinders to provide a variety of mechanical motions.

161

motion into either oscillating or rotary motion. By selecting the correct linkage, a designer can increase or decrease the effective leverage or length of a cylinder stroke.

In other applications, it may be necessary to obtain positional advance motions. Cams, cogs, gears, segments, and screw drives may be employed. First, the designer must determine how a given force is to be transformed into a precise pattern of motion. Then, the linkage that is most practical, safe, efficient, and compatible with the particular application can be selected.

Designers of hydraulically equipped machines or equipment need to analyze an entire system, including the magnitude and direction of the force, the distance through which the force will move (stroke), the time it will take the force to act, and the sequence of operations.

Example 8.3 ▶ Suppose it is necessary to move a force of 3500 lb by a verticle distance of 6 ft in 6 s, and lower the load in 4 s. Your lab facility has a hydraulic hoist with a 4 in. diameter cylinder and a 2 in. diameter rod. Calculate the pump capacity and pressure rating to achieve the necessary conditions.

Solution: First compute the cap-end and rod-end areas.

$$\text{cap end} = A = 0.7854D^2 = 0.7854(4)^2 = 12.56 \text{ in.}^2$$

$$\text{rod end} = 12.56 - 0.7854(2)^2$$

$$\text{annular area} = 12.56 - 3.14 = 9.42 \text{ in.}^2$$

$$p = \frac{F}{A} = \frac{3500}{12.56} = 278.66 \text{ psi}$$

$$\text{time to extend piston (min)} = \frac{\text{cylinder displacement (in.}^3)}{\text{pump displacement in.}^3/\text{min})}$$

If the extension stroke takes 4 min, we can compute the gpm of the pump required to produce the necessary speed of the extension stroke.

$$\frac{6}{60} = \frac{\text{area} \times \text{stroke (in.}^3)}{\text{pump volumetric displacement (in.}^3/\text{min})}$$

$$\frac{6}{60}V_\text{D} = \frac{12.56 \times 6 \times 12}{1}$$

$$V_\text{D} = \frac{12.56 \times 6 \times 12 \times 60}{6}$$

$$V_\text{D} = 9043.2 \text{ in.}^3/\text{min}$$

$$1 \text{ gpm} = 231 \text{ in.}^3/\text{min}$$

To convert the pump volumetric displacement to gpm

$$V_{\text{D}} = \frac{9043.2}{231} = 39.15 \text{ gpm}$$

On the retraction stroke

$$V_{\text{D}} = \frac{9.42 \times 6 \times 12 \times 60}{4} = 10{,}173.6$$

$$\text{gpm} = \frac{10{,}173.6}{231} = 44.04$$

Therefore, a pump with a displacement of 45 gpm and a 500 psi rating would be adequate to provide the necessary margin of safety to raise the load of 3500 lb in 6 s. Some degree of metering would be necessary to maintain the precise velocity per minute. ◀

8.7 HYDRAULIC MOTORS

Hydraulic motors convert fluid energy into a mechanical motion or force (Figure 8.14). They are also called *rotary actuators*. The elements of a hydraulic

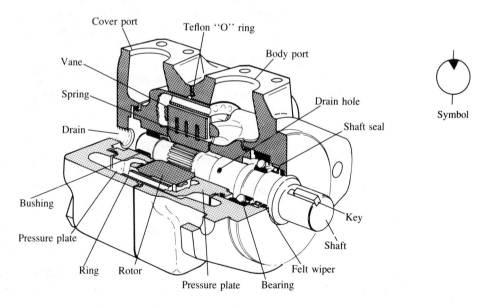

FIGURE 8.14
Components of a fluid motor.

motor are pushed by the pressurized fluid. In this manner, the hydraulic motor converts fluid pressure energy into a torque and produces continuous rotary motion of the motor shaft.

8.8 OPERATION OF HYDRAULIC MOTORS

All hydraulic motors, gears, vanes, and pistons have a driving surface area subject to pressure differential. The surface area of each element is connected mechanically to an output shaft. Hydraulic motors are designed to be unidirectional (operate in one direction) or bidirectional (operate in both directions). By reversing the direction of flow it is possible to reverse the rotation of the motor shaft.

The principal ratings of a hydraulic motor are pressure, displacement, and torque. Torque and pressure ratings of a motor indicate how much load the motor can provide. Displacement refers to the amount of fluid required to turn the motor output shaft one revolution. The units of measure of motor displacement are cubic inches per revolution (in.^{3}/rev) or cubic centimeters per revolution (cm^3/rev). Displacement of hydraulic motors may be fixed or variable. A fixed-displacement motor provides constant torque; speed is varied by controlling the amount of input flow into the motor. A variable-displacement motor provides variable torque and variable speed. With input flow and pressure held constant, the torque speed ratio can be varied to meet the load requirements by varying the displacement.

It is possible to compute the volumetric delivery needed to attain a given motor speed if the displacement of the motor is identified. For example, if the motor speed of 1000 rpm and a displacement rating of 2.31 in.^{3}/rev are used, the equation would be

$$\text{gpm} = \frac{\text{speed (rpm)} \times \text{displacement (in.}^3\text{/rev)}}{231}$$

Substituting the given values in the example

$$\text{gpm} = \frac{1000 \times 2.31}{231} = 10$$

If the displacement of the motor and the gpm delivery of the pump are known, the drive speed of the motor in revolutions per minute (rpm) can be computed as follows

$$\text{rpm} = \frac{\text{gpm} \times 231}{\text{displacement (in.}^3\text{/rev)}}$$

Substituting the given values into the previous equation

$$\text{rpm} = \frac{10 \times 231}{2.31} = 1000$$

Torque output is a rotary thrust. Motor torque is expressed in lb·in. or lb·ft. Torque capacity is proportional to both the pressure and volumetric displacement.

The equation for computing torque (lb·in.) is

$$T = \frac{\text{pressure} \times \text{displacement (in.}^3\text{/rev)}}{2\pi} = \frac{pV_D}{6.28}$$

If a motor has a displacement of 5 in.³/rev and a working pressure of 1000 psi, the motor torque would be

$$T = \frac{(1000)\,(5)}{6.28} = 796 \text{ lb·in.}$$

This example shows that the motor torque increases when either the pressure or the displacement increases. In a variable-displacement motor it would be possible to increase the displacement. However, as displacement is increased, the speed of the motor is decreased in proportion to the gain in torque.

8.9 HORSEPOWER

The horsepower (*HP*) output for a hydraulic motor can be expressed mathematically as

$$HP = \frac{\text{torque (lb·in.)} \times \text{rpm}}{63,025} \quad (33,000 \times 12)$$

or

$$HP = \frac{T \times \text{rpm}}{63,025}$$

For example, if in the previous example the motor produced 795 lb·in of torque and the output motor shaft made 1800 rpm, the horsepower output of the motor would be

$$HP = \frac{796 \times 1800}{63,025} = 22.73$$

The hydraulic horsepower can be calculated, if the pressure (psi) and the flow rate (gpm) are known. For example, if in the previous examples the motor displacement is 5 in.^{3}/rev, the gpm for 1800 revolutions will be $1800 \times 5 = 9000$ in.^{3}/min. This flow rate represents $9000/231 = 38.96$ gpm.

$$HP = \frac{\text{gpm} \times \text{psi} \times 0.583}{1000}$$

$$HP = \frac{39 \times 1000 \times 0.583}{1000}$$

$$HP = 22.73$$

8.10 OPERATION OF GEAR MOTORS

External gear motors (Figure 8.15) consist of a pair of matched gears enclosed in one housing. Both gears are driven in a motor by pressurized fluid, but only one gear is connected mechanically to the output shaft. The operation of an external gear motor is essentially the reverse of that of a gear pump.

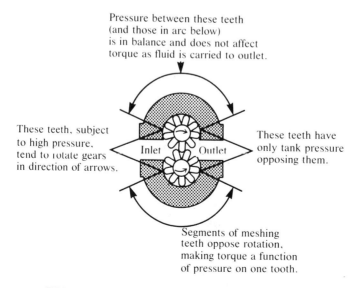

FIGURE 8.15
Torque development of a gear motor.

Pressurized fluid enters the motor housing at the inlet port and is prevented from flowing directly to the outlet port because of the meshed teeth. As a result, the fluid forces the gears to rotate following the path of least resistance around the periphery of the housing (see Figure 8.15).

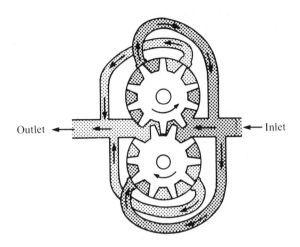

FIGURE 8.16
Balanced gear motor.

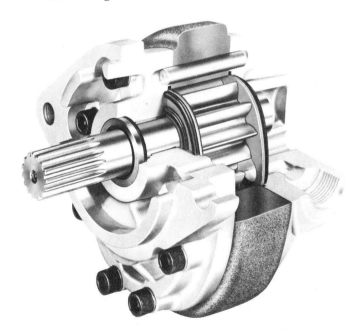

FIGURE 8.17
External gear motor. (Courtesy Webster Electric Company, Inc.)

After traveling around the peripheral channel of the housing, the fluid exhausts to the reservoir. The direction of rotation of a hydraulic motor can be reversed by reversing the direction of fluid flow through the ports. The volumetric displacement of a gear motor is fixed, as is also the case with gear pumps.

Generally, gear motors are not balanced with respect to pressure loads. The high pressure at the inlet, with the opposed low pressure at the outlet, produces a large side load on the shaft and bearings. Some gear motor manufacturers balance this side load by internal cored passages and multiple ports that distribute identical opposing pressure conditions (180° apart as illustrated in Figure 8.16). External gear motors have wear plates on the sides of gears to keep the gears from moving axially and help control leakage and lubrication.

External-spur gear motors are normally limited to 2000 psi operating pressures and provide a top speed of 2400 rpm. External gear motors are available with a maximum flow capacity of 150 gpm. External gear motors generally have an overall efficiency of approximately 70 to 75 percent, compared to 85 to 95 percent for piston motors. The main features of a hydraulic gear motor are its simple design, lower cost, and ease of maintenance (Figure 8.17).

8.11 INTERNAL GEAR MOTORS

This type of hydraulic motor consists of an inner and outer gear set with the inner gear mounted on the output motor shaft. The inner gear has one less

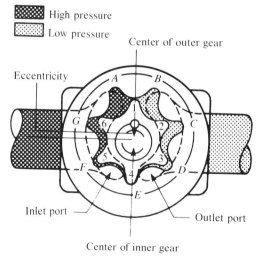

FIGURE 8.18
Internal gear motor.

(a)

(b)

FIGURE 8.19
(a) Major parts of a gerotor (internal gear) motor. (Courtesy Parker Fluidpower) (b) Main elements of gerotor hydraulic motor. (Courtesy Eaton Corporation)

tooth than the outer gear. The geometric shape of the teeth is such that all teeth of the inner gear are in contact at all times with some portion of the outer gear.

While the center of the inner gear coincides with the center of the output shaft, the outer gear is offset according to the design eccentricity.

As pressurized fluid enters the inlet port (because the inner gear has one less tooth than the outer gear), a pocket is formed between teeth 6 and 1 and the outer gear socket (Figure 8.18). As teeth 6 and 1 rotate past the maximum area of the kidney-shaped inlet port, fluid flow is sealed off. The tips of inner gear 6 and 1 provide the seal between the gear set. As the pair of inner and outer gears continue to rotate, a new pocket is sealed off by the next adjacent inner gear teeth with the outer gear. Meanwhile, the fluid is steadily exhausted through the outlet port. The kidney-shaped ports allow smooth, uniform fluid flow and rotation of the motor output shaft.

Internal gear motors generally operate at a higher pressure, higher speed, and larger displacements than the external gear motors (Figure 8.19). Remember, the larger the gear or the higher the system pressure, the more torque will be developed at the shaft.

8.12 VANE MOTORS

Vane motors develop torque by the fluid pressure acting on the exposed vanes, which slide out of a rotor connected to the motor output shaft (Figure 8.20). A balanced vane motor consists of a cam ring, slotted rotor, multiple vanes, and a port plate with inlet and outlet ports opposite each other. Both inlet ports are connected as are the outlet ports, so that each set can be served by one inlet or

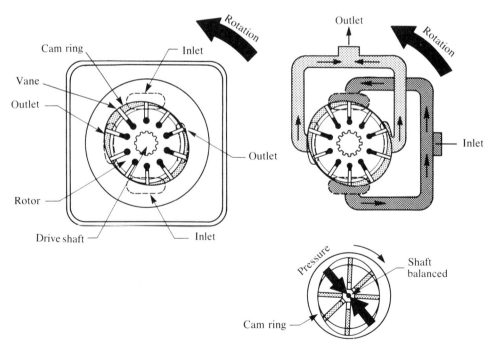

FIGURE 8.20
Balanced vane motor design.

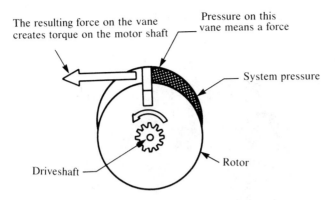

The resulting force on the vane creates torque on the motor shaft

Pressure on this vane means a force

System pressure

Driveshaft

Rotor

FIGURE 8.21
Principle of operation of a vane motor.

one outlet in the motor housing. The pressure areas are diametrically opposed so as to balance the hydraulic pressure forces.

Mechanical springs are used in some motor designs to hold the vanes against the cam ring at low initial speeds. Still other designs have pressure-loaded vanes (Figure 8.21). The rotating group of a vane motor includes an integral cartridge assembly consisting of vanes, rotor, and a cam ring sandwiched between two port plates. This makes it relatively easy to service a vane motor.

Vane motors (Figure 8.22) provide good operating efficiencies of 75 to 85 percent, but not as high as piston motors. Normally they have a higher tolerance for fluid contamination than the piston motors. However, they do have a shorter service life than a piston motor and also have limited low-speed capability. Vane motors have two critical seal points: the points of contact between the vanes and the cam ring, and the contacting surfaces between the cartridge assembly and the pressure plates. Most reversible hydraulic motors require an external drain line to conduct leakage oil back to the reservoir.

8.13 PISTON MOTORS

Piston motors are designed to function as fixed- or variable-displacement models. The two types of rotary piston motors are the axial and the radial. The pistons in an axial piston motor reciprocate in a plane parallel to the output shaft. The pistons in a radial piston motor reciprocate perpendicular to the output shaft.

The operation of a piston motor is based on its ability to generate torque by pressure acting on the ends of pistons reciprocating in a cylinder block.

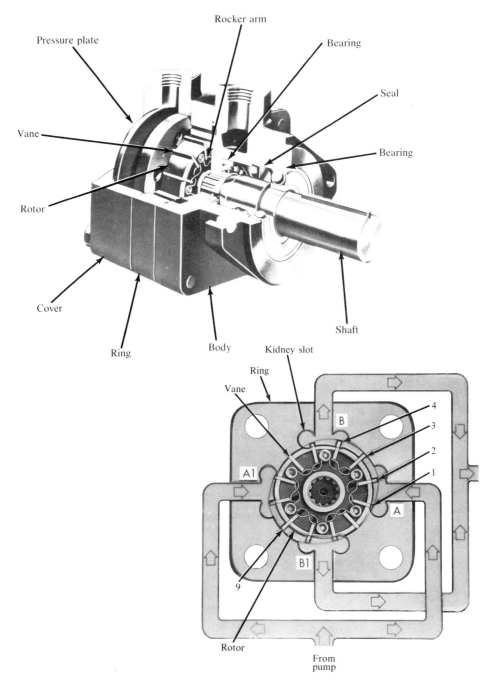

FIGURE 8.22
(a) Vane motor. (Courtesy Vickers, Incorporated) (b) Vane motor. (Courtesy Vickers, Incorporated)

Figure 8.23 illustrates the in-line piston motor. In this design type, the motor drive shaft and cylinder block are centered on the same axis. Pressure at the ends of the multiple pistons causes a sequential reaction against a tilted swashplate and rotates the cylinder block and motor shaft. Torque is proportional to the area of the pistons, and is a function of the angle at which the swashplate is pivoted.

In the fixed-displacement motor, the swashplate remains at a fixed angle. In the variable-displacement model, the swashplate is mounted in a pivoted yoke, which allows the angle to be altered by a variety of means. With variable-displacement motors, increasing the swashplate angle increases torque capability but reduces the driveshaft speed. Reducing the swashplate angle also

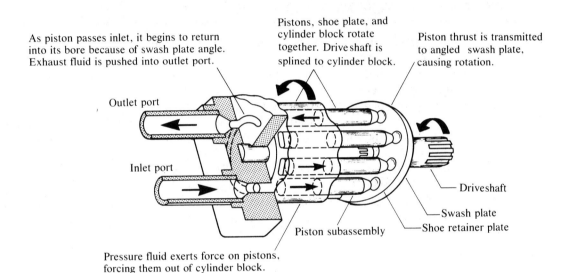

As piston passes inlet, it begins to return into its bore because of swash plate angle. Exhaust fluid is pushed into outlet port.

Pistons, shoe plate, and cylinder block rotate together. Drive shaft is splined to cylinder block.

Piston thrust is transmitted to angled swash plate, causing rotation.

Outlet port

Inlet port

Driveshaft

Swash plate

Shoe retainer plate

Piston subassembly

Pressure fluid exerts force on pistons, forcing them out of cylinder block.

FIGURE 8.23
In-line piston motor.

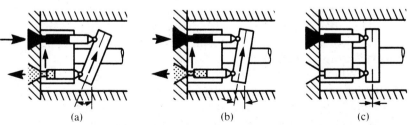

(a)

(b)

(c)

Maximum swash plate angle (maximum displacement)

Decreased swash plate angle (partial displacement)

Minimum swash plate angle (zero displacement)

FIGURE 8.24
Varying displacement of an in-line piston pump.

reduces the torque but increases the driveshaft speed if the same pressure is maintained.

The displacement of a piston motor is determined by the distance the pistons are reciprocated in the cyclinder barrel and by the size of the pistons (Figure 8.24). The swashplate angle controls the distance in an axial piston motor. With a small swashplate angle, the pistons have a short stroke within the cylinder barrel. Therefore, for a given pump displacement, the motor speed would increase as the swashplate angle is reduced. Therefore, it is possible to alter both the speed and torque of a motor by varying the swashplate angle.

8.14 BENT-AXIS PISTON MOTORS

Bent-axis piston motors are available in fixed- or variable-displacement models. The operation of a bent-axis piston motor is similar to the in-line piston motor except that the thrust of the pistons is transferred against the driveshaft (Figure 8.25). The torque of this piston motor is at its maximum when the displacement is at its maximum. Conversely, speed is at its minimum.

The angle generally varies from 7½° to 30°. Some piston motors have manual controls whereas others include the more sophisticated servo-valve controls.

Piston motors are the most generally efficient of the three basic types of motors (85 to 95 percent efficiency) and operate at the highest speeds and pressures. Operating speeds of 12,000 rpm and pressures of 5000 psi can be attained with piston motors. The maintance of this efficient motor is more critical and difficult to perform.

8.15 RADIAL PISTON MOTOR

This piston motor has a cylinder barrel with an attached output shaft to transmit the force imparted to the pistons (Figure 8.26). The cylinder bore has an odd number of radial bores with precision-fitted pistons. In a variable-displacement model, the thrust ring is adjustable. In a fixed-displacement model, the thrust ring is stationary.

When oil enters the cylinder bore, the piston is forced against the thrust ring, imparting a tangential force to the cylinder barrel and shaft, causing the assembly to rotate. Each piston is pushed inward (Figure 8.26) by the thrust ring once it reaches the outlet port, thus pushing the fluid to the reservoir. Radial piston motors are also equipped with an external drain to prevent pressurizing the cylinder chamber. They usually contain 3, 5, 7, 9, or more pistons, depending on the size of the unit.

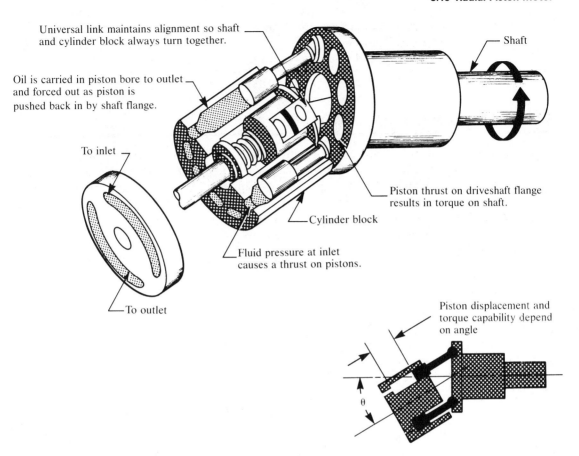

Universal link maintains alignment so shaft and cylinder block always turn together.

Shaft

Oil is carried in piston bore to outlet and forced out as piston is pushed back in by shaft flange.

To inlet

Piston thrust on driveshaft flange results in torque on shaft.

Cylinder block

To outlet

Fluid pressure at inlet causes a thrust on pistons.

Piston displacement and torque capability depend on angle

θ

FIGURE 8.25
Bent-axis piston motor.

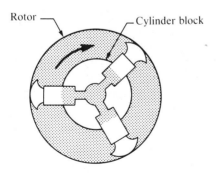

Rotor

Cylinder block

FIGURE 8.26
Radial-piston hydraulic motor.

Rotary piston motors are generally compact units, providing high torque and high acceleration, with excellent life expectancy when operated within load limitations and under good maintenance practices.

8.16 SIZING A HYDRAULIC MOTOR

The maximum horsepower produced by a motor is reached when it is operating at the maximum system pressure and at the maximum shaft speed. For example, if a hydraulic motor application required 5 hp at 3000 rpm, with an available supply pressure of 3000 psi and a 100 psi return-line pressure loss, the effective pressure differential will be 2900 psi.

The theoretical torque is calculated as follows

$$T = \frac{63{,}025 \times HP}{\text{speed}}$$

$$T = \frac{63{,}025 \times 5}{3000}$$

$$T = 105.04 \text{ lb} \cdot \text{in.}$$

Motor displacement is calculated as follows:

$$D = \frac{2\pi T}{p e_{\text{m}}}$$

where D = displacement, in.^{3}/rev
T = torque, lb $\cdot$ in.
p = differential pressure, psi
e_{m} = mechanical efficiency, % , 85

If the mechanical efficiency is 85% then

$$D = \frac{6.28 \times 105}{2900 \times .85}$$

$$D = 0.2675 \text{ in.}^3/\text{rev}$$

The fluid flow required is

$$Q = \frac{DN}{231 e_{\text{v}}}$$

where Q = flow, gpm
         D = displacement, in.^{3}/rev
         N = speed, rpm
         e_v = volumetric efficiency, %

If the volumetric efficiency of this particular motor is 94%, we can compute gpm as follows

$$Q = \frac{0.267 \times 3000}{231 \times .94}$$

$$Q = 3.695 \text{ gpm}$$

Em• Ev=Eoverall

Thus, any pressure at the outlet port will reduce the motor torque output.

Summary

Hydraulic actuators provide linear or rotary motions by converting pressurized fluid energy into mechanical motion. Cylinders should be properly mounted and aligned to avoid unnecessary stresses on the rod and bearings. The proper selection, installation, and maintenance of cylinders and motors will extend their life expectancy.

Cylinders are manufactured in a variety of types, mounting arrangements, and sizes to accommodate a wide range of applications. The double-acting cylinder is used very widely in numerous applications. Cushioning of cylinders protects a cylinder from stresses by decelerating the stroke at its extremes.

Hydraulic motors provide rotary motion and are available in gear, vane, and piston types. The piston motor, which is the most efficient, is manufactured in fixed- and variable-displacement models. The variable-displacement motor may be controlled by changing the angle of the yoke or stroking ring by several means. They can be mechanical, electric, hydraulic, or pneumatic.

It is possible to get somewhat less energy out of an actuator in relation to the input energy for a given interval of time depending on the efficiency of a unit. If the torque of a motor increases, its speed decreases. If the load on a cylinder increases, its speed is also reduced.

A very large number of necessary motions to accomplish work can be created by connecting actuators to cams, gears, belts, rack and pinion, and other mechanisms to achieve the desired motions.

Sequencing, positioning, and controlling speeds of actuators with increasingly sophisticated control valves are keeping hydraulic actuator applications more at the forefront in diversified industries.

Questions and Problems

1. Describe the operation of a double-acting cylinder.
2. What is the purpose of a cylinder cushion?
3. Where would a telescoping cylinder be used?
4. With a 20 gpm pump supplying hydraulic oil to a 2 in. diameter, double-acting hydraulic cylinder with a 1 in. diameter rod and a load of 1000 lb (extension and retraction strokes), compute the following:
   a. Pressure during the extension stroke.
   b. Piston velocity during the extension stroke.
   c. Pressure during the retracting stroke.
   d. Piston velocity during the retraction stroke.
5. Explain what a balanced gear motor is.
6. What are the main advantages of gear motors?
7. Why must hydraulic motors be externally drained?
8. A hydraulic motor has a volumetric displacement of 2.31 in.^{3}/rev. If it has a pressure rating of 3000 psi and is connected to a 15 gpm pump compute the motor's:
   a. speed
   b. torque
   c. horsepower
9. Given a fluid motor with a displacement of 6.0 in.^{3}/rev and a required rpm of 1000, determine the gpm rating of the pump.
10. Describe those circuit requirements that would be better served by a hydraulic piston motor than a gear motor.

Suggested Learning Activities

1. Disassemble a double-acting cylinder and observe the method of cushioning the cylinder. Draw a sketch to illustrate how the piston is cushioned.
2. Disassemble a hydraulic gear motor and determine whether it is a balanced gear motor.
3. Disassemble a hydraulic vane motor and determine how the vanes are held against the cam ring.
4. Disassemble a hydraulic piston pump and identify the number and size of the piston.
5. Locate a variable-displacement, bent-axis piston pump on any piece of equipment. Determine how the adjustment of the angle of the swashplate is controlled.

Suggested Readings

Basal, P. R., Jr. (ed.) *Mobile Hydraulics Manual,* Sperry Rand Corporation, Troy, Michigan, 1967, pp. 61–79.

Esposito, Anthony. *Fluid Power with Applications.* Englewood Cliffs, N.J.: Prentice-Hall, 1980, pp. 177–206.

Henke, Russell W. "Fluid Power Systems and Circuits." *Hydraulics and Pneumatics Magazine,* Cleveland, Oh., 1983, pp. 256–71.

Sullivan, James A. *Fluid Power: Theory and Applications,* 2nd ed. Reston, Va.: Reston, 1982.

Wolansky, William; Nagohosian, John; and Henke, Russell W. *Fundamentals of Fluid Power.* Boston: Houghton Mifflin, 1977, pp. 171–203.

Yeaple, Frank. *Fluid Power Design Handbook.* New York: Dekker, 1984, pp. 78–136.

CHAPTER 9

System Design and Basic Circuits

Introduction

Previous chapters presented the theory and operation of the components that are essential to constructing hydraulic circuits. This chapter enables you to apply the theory of fluid mechanics and your knowledge of hydraulic components to the practical design of hydraulic circuits. Although you now should be able to accomplish basic circuit design, remember that only after considerable experience and practice can highly successful system designs be achieved.

Mastery of fluid mechanics theory, plus comprehension of a sound knowledge of the principles of hydraulic components operation and an awareness of interface compatibility of components will enable you to design and fabricate basic circuits (Figure 9.1).

Key Terms

Analysis: Analysis of a circuit implies a scientific study of energy transfer within a system to perform a specific task.

Closed-centered Circuit: An open-loop system in which the pump or accumulator source is blocked by the directional control valve when in neutral position.

Closed-loop System: A system that uses feedback to produce a self-adjusting or self-regulating system.

Design: Design of a circuit implies a synthesis of an energy transfer system to perform a specific task.

Feedback: The technique of using a transducer to generate a signal proportional to, or representative of, the state of the system output variable. Next, compare this (feedback) signal to a command signal and adjust the system to make the output conform as closely as possible to the desired level.

Load: The complete performance characteristic of a machine expressed in load reaction (force) magnitude, sense, and velocity for one complete machine cycle.

Hydraulic System: A system of interconnected, selected components and conductors to provide required work functions by the transmission, control, and utilization of a pressurized fluid. A fluid power system may be composed of many circuits.

9.1 CRITERIA FOR DESIGNING CIRCUITS

A *hydraulic circuit* is an arrangement of interconnected components selected to achieve desired work output. A *hydraulic system* is the interconnection of selected components to provide required work functions by the transmission, control, and utilization of a pressurized fluid (Figure 9.2).

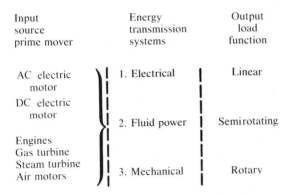

FIGURE 9.1
Block diagram of energy transmission systems.

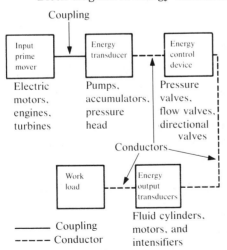

FIGURE 9.2
Block diagram illustrating an energy system.

Systems design complexity can range from designing the circuit of a hydraulic jack to the programmable controller remotely raising or lowering an automobile. It can be as complex as a hydraulic system that lands a large aircraft on automatic pilot control. The nature of the task will dictate the range of system parameters that must be met by the particular circuit or combination of circuits. There are many factors and criteria that can be identified as useful guides for selecting hydraulic circuits. Several of the factors would include the following

1. The system must be designed to operate safely. This includes protective devices for operators and service personnel as well as protection of the system components against excessive pressures, temperatures, flow rates, and fire hazards. Emergency shutdown features and power failure locks should also be considered.
2. The system must be functional, reliable, and cost effective. System components should be compatible, provide an adequate margin of safety to withstand operational hazards, exceed the required circuit specifications, facilitate maintenance practices, and be reasonable in cost.
3. The system must be energy efficient. Most hydraulic components are relatively efficient. However, as these are interconnected, efficiency drops off rather rapidly through thermal, mechanical, functional, and line transfer losses, plus pressure drops across valves.

The system design should be simple, compact, and functional. Use of standardized components with the fewest connectors, good contamination control, and good fabrication techniques can enhance the efficiency of a hydraulic system.

It is possible to underdesign or overdesign a hydraulic system. This could lead to premature failure or unnecessary cost overrun. Optimum designs require a thorough knowledge of the theory and components, plus extensive successful experience in solving circuit design problems. Analysis of the task, determination of specifications, selection of components, fabrication, and testing of the circuit under real operating conditions will facilitate acceptable circuit design solutions. Remember that many cycles in the design procedure and testing are necessary to achieve an efficient and safe design circuit.

9.2 CRITERIA FOR SELECTING CIRCUITS

Circuit design implies problem-solving. Therefore, an analytical technique comprised of defining the problem, calculating specification requirements, examining alternate solutions (and their limitations), and verifying the adequacy of the solution improve the likelihood for a more acceptable circuit design. A circuit design check list might include

- Specification of the task to be accomplished. Next, determine:
- The time-per-work cycle.
- The flow-rate distribution over the work cycle.
- The pressure distribution over the work cycle.
- The horsepower demand pattern.
- The cost of the circuit.
- The adequacy of the safety features.

9.3 CIRCUIT DESIGN

Fluid power circuits from a design standpoint consist of four critical sections (Figure 9.3). Section I represents the energy input available to perform work. Energy is transferred to the load across the mechanical interface involving a cylinder in this circuit. Section II includes the control where fluid modulation and directional switching occur.

The energy output portion of a hydraulic circuit is indicated in Section III of Figure 9.3. This section includes the pump, which is coupled to a prime mover. Due to frictional and heat losses, the energy at the output must exceed that of the required energy input to the load. Pump efficiencies were discussed in Chapter 6. Section IV includes the auxiliary components such as conductors, fittings, manifolds, reservoirs, filters, gauges, and flow meters.

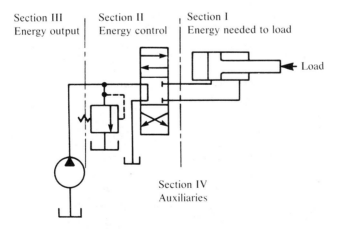

FIGURE 9.3
Hydraulic circuit divided into four essential sections.

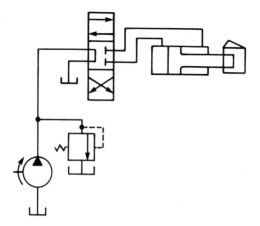

FIGURE 9.4
Simple reciprocating piston rod circuit for splitting a log.

All four sections are critical to the design of a functional, safe, and efficient circuit. When drawing the diagram of a circuit, you normally begin with a pump and end with the actuator. In designing a hydraulic circuit, the reverse pattern holds true. The designer begins with the load or task, and works backward through the actuator, controls, pump, and ultimately, the prime mover. Auxiliaries are also important. For example, piping must be sized to permit laminar flow at a given rate for a known pump displacement. It may be helpful to go through the steps necessary to design a very simple, reciprocating piston circuit used to split a log (Figure 9.4).

Example 9.1 ▶ Suppose these are the known task specifications for a log splitter circuit.

- Energy input load requirements for splitter = 20,000 lb.
- Stroke of piston rod (maximum log length) = 20 in.
- Given time interval for extension stroke = 2–4 s.
- Safety factor = 1.3–1.5.

Economic and maintenance considerations would dictate the choice of a gear pump for this circuit (see Table 6.1 in Chapter 6). It indicates the pressure rating could vary from 2000 to 3000 psi. This requirement dictates the system pressure.

Given this information, it is possible to calculate the ranges of cylinder diameters (thus, oil flow rate) and the power requirements of the system, including allowance for power efficiency losses. Once these calculations are made, it will be possible to choose a suitable prime mover, likely a diesel engine. Let's go through the necessary computational steps for designing a very simple, reciprocating piston circuit.

Solution:

1. Compute the diameter ranges for the 2000 and 3000 psi limits.

$$\text{force} \times \text{SF} = \frac{D_P{}^2}{4} \text{ (pressure)}$$

where   force = 20,000 lb

SF = safety factor ratio

D_P = diameter of piston

Thus, the maximum D_P is when SF = 1.5 and p = 2000 psi.

$$\text{max. } D_P = \frac{20{,}000 \times 1.5}{(0.7854)\,(2000)} = 4.37 \text{ in.}$$

We can select the nearest functional diameter piston of $4\frac{3}{8}$ in. or 4.375 in.

$$\text{min. } D_P = \frac{20{,}000 \times 1.3}{(0.7854)\,(3000)} = 3.32 \text{ in.}$$

For this limit we can select $3\frac{3}{8}$ in. or 3.375 in.

2. Now compute the maximum and minimum flow rates and power requirements. Assume an overall efficiency from Table 6.1 to be 80 percent.

Maximum conditions occur for minimum time of piston stroke of 2 s and maximum piston diameter.

Thus, the maximum flow rate

$$\text{volume (gpm)} = \frac{A \times L \times 60}{231 \times \text{(time in seconds)}}$$

where A = area

L = length of stroke

$$\text{gpm} = \frac{0.7854\,(4.375)^2 \times 20 \times 60}{231 \times 2} = 39.04$$

The maximum power would result when the maximum pressure demanded was 3000 psi.

$$H_P = \frac{\text{gpm} \times \text{psi}}{1714 \times \text{pump efficiency}} = \frac{39.04 \times 3000}{1714 \times 0.80} = 85.41$$

Thus, a diesel engine would have to deliver up to 86 hp. ◄

There are very many considerations in designing a hydraulic system. From the beginning, the magnitude of load to be moved with controlled time cycle will be critical. A study of the loads, sequence, and time intervals will help establish the major choices of an actuator, controls, pump, and capacity of the prime mover.

At the same time, a system designer must be familiar with components. Guidelines for selecting hydraulic components include

1. Job application considerations.
   a. Type of applications:
      - Aerospace
      - Industrial
      - Marine
      - Mobile
   b. Environmental conditions:
      - Temperature ranges
      - Atmospheric cleanliness
2. Prime mover characteristics considerations.
   a. Type of prime mover available:
      - Speed (rmp)
      - Speed variation
   b. Mounting of prime mover.
   c. Drive methods:
      - Chain versus belt or shaft coupling
   d. Power output available.
   e. Economics (initial cost and maintenance).
3. Type of fluid considerations.
   a. Mineral use:
      - Water glycol
      - Emulsions
      - Phosphate esters
      - Other synthetics
   b. Seal and wiper compatibility.
4. Economic and customer considerations.
   a. Availability:
      - Customer acceptance
      - Initial and replacement costs
      - Ready compatibility with other components
   b. Safety (are sufficient safety devices included in the design?):
      - Check valves
      - Safety controls
      - Instructions for maintenance workers

The circuit designer must be able to design a hydraulic system that accomplishes the required task within predetermined specifications. Components must be selected that will function reliably, safely, and economically in the system over a designated life expectancy with quality maintenance. For these reasons, design takes considerable knowledge of theory plus practical experience and a thorough acquaintance with components.

The more complex the operations to be performed, the more critical the limits of control. The more complex the components included in the system, the more difficult are the design solutions. Yet, a systems approach allows sequential procedures to compute, organize, and assemble efficient circuits.

It is relatively easy to design a simple, linear reciprocating, single-actuator circuit. For example, such a circuit is used to split a log with a hydraulic cylinder. The log splitter confronts a resistive load; that is, the load reaction on the output device opposes the motion of the actuator.

9.4 OPEN-LOOP CIRCUITS

The three functions performed in an open-loop circuit are: directional controls regulate the direction of the distribution of energy, flow controls regulate the rate at which energy is transferred by adjusting flow rate in a circuit or branch of circuits, and pressure controls regulate energy transfer by adjusting pressure level or by using a specific pressure level as a signal to initiate a secondary action (Figure 9.5). An open-loop circuit has no feedback mechanism to monitor system output and compare it to an input or command signal as does a closed-loop system. The two types of open-loop circuits are constant flow and demand flow.

9.5 CONSTANT FLOW CIRCUITS

In a constant flow circuit (Figure 9.5), the directional control valve bypasses fluid to the reservoir when the valve is in the neutral (center) position. This unloads the pump to the reservoir. This schematic circuit includes a fixed-

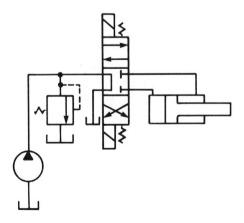

FIGURE 9.5
Typical open-loop circuit.

displacement pump protected by a relief valve. Moving the spring-centered directional valve spool starts energy transfer from essentially point zero to ultimately reach the relief valve setting.

Constant flow circuits are economical. This circuit can operate at very minimal pressure during the neutral position, yet provide the necessary flow when the directional valve is shifted.

9.6 DEMAND FLOW CIRCUITS

The circuit in Figure 9.6 includes a fixed-displacement pump, an accumulator, and an unloading valve. In this circuit all ports of the directional valve are blocked in the neutral position.

Unlike the constant flow circuit, where pressure starts from virtually zero and builds up as the spool is shifted, in a demand flow circuit the energy starts from the maximum pressure setting of the system. In the neutral position, there is a greater concern for internal leakage of the directional valve than in a constant flow. This is because the spool is holding all ports closed against the full system pressure. A demand flow circuit can also be assembled using a pressure-compensated pump to replace the fixed-displacement pump, an accumulator, and an unloading valve.

Energy transfer circuits primarily provide directional control. They are used where it is desirable to regulate acceleration, velocity, or deceleration of a linear or rotary actuator. Lift mechanisms, steering mechanisms, and cargo handling mechanisms all use directional and flow control. Where it is essential to have positive and instant control, the demand flow circuit will provide this response as soon as the directional valve spool is shifted.

9.7 CLOSED-LOOP SYSTEMS

A closed-loop system includes a feedback mechanism, which continually monitors systems output and compares it to an input or command signal. The feedback mechanism (servo valve) compares the system's output to an input or command signal. If there is a difference between the input command signal and the feedback signal, the output is adjusted automatically to match the programmed command requirements.

The electrohydraulic servo system (Figure 9.7) is a feedback system in which the output is a mechanical position that interacts with the servo-valve directional control valve. A closed-loop system can also include constant flow circuits or demand flow circuits. The feedback mechanism will be sensing the load of a linear actuator or torque of a rotary actuator employing an electrohydraulic servo-valve. While constant flow circuits in closed-loop systems commonly

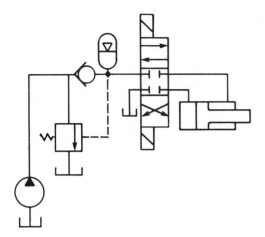

FIGURE 9.6
Demand flow circuit. All ports are blocked in neutral position.

employ fixed-displacement pumps, the demand flow circuit employs variable pressure-compensated pumps.

It is helpful to note that an open-loop circuit does not contain a feedback mechanism as is the case with a closed-loop circuit. Also, in a constant flow circuit, fluid bypasses to the reservoir over the spool of the directional valve at virtually zero pressure. In a demand flow circuit, all the ports are blocked all the time in the neutral position of the directional valve, thereby maintaining system pressure (Figure 9.8).

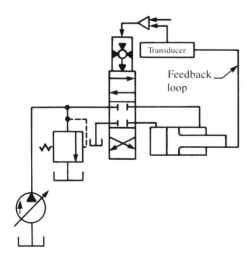

FIGURE 9.7
Closed-loop circuit.

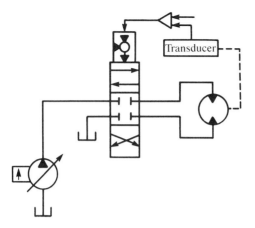

FIGURE 9.8
Closed-loop system, demand flow, open circuit.

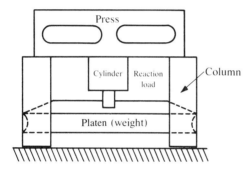

FIGURE 9.9
Overrunning load on press augments motion of the actuator in the same direction.

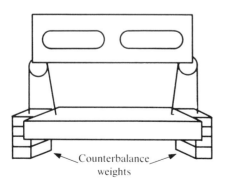

FIGURE 9.10
Counterbalance weights to overcome the weight of the platen and die mold.

The signal between the rotary hydraulic motor and the variable pressure-compensated pump will be continuously monitored. With the directional valve spool in the neutral position, all ports are blocked with system pressure being maintained, and the pump will deliver minimum flow.

The type of application being used will influence the circuit designer's choice of an open-loop or closed-loop system, and determine whether a constant flow or a demand flow circuit would be more appropriate.

9.8 CIRCUIT FOR OVERRUNNING LOAD

The purpose of every hydraulic circuit is to transfer energy to an output device or actuator so it can do useful work, such as moving or restraining a load, or exerting a force.

Many circuit applications are designed to move resistive loads. This is a load that opposes the motion of the actuator, such as lifting a loaded aggregate box on a dump truck. Other applications encounter an overrunning load, one in which the load reaction is in the same direction as the actuator motion. For example, a sheet metal press mold platen used to stamp an automotive panel may weigh 2 tons (Figure 9.9). On the extension stroke of the actuator, the added weight of the platen augments the downward force and may tend to cause an overrun of the extension if control is not provided.

It is possible to use mechanical means to counterbalance the weight of the platen and die mold (Figure 9.10). However, each time the size and weight of the die mold are changed, the balance will be altered.

9.9 COUNTERBALANCE CIRCUIT

The primary function of a counterbalance valve is to restrict fluid flow from the primary to the secondary port (Figure 9.11) and to maintain a pressure level sufficient to balance a load being held by an actuator. The counterbalance valve is a normally-closed, internally drained valve. It operates on the principle that fluid is trapped under pressure until pilot pressure (either direct or remote) overcomes the spring force setting in the valve and the spool moves up to throttle return flow through the valve to the reservoir (Figure 9.11). An internal check valve permits free flow around the piston when the directional valve is shifted to raise the load or press platen.

Counterbalance valves are used in such applications as loaders, lift trucks, vertical presses, and similar circuit applications where forces have to be balanced hydraulically while the pump is idling. The counterbalance valve is set to open slightly above the pressure required to hold the piston up. This circuit

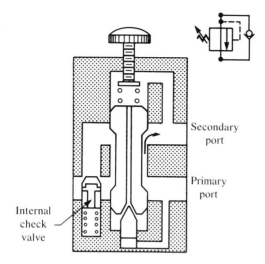

FIGURE 9.11
Counterbalance valve.

permits the cylinder to be forced downward when pressure is applied on the blind end.

It is much more convenient to use a hydraulic counterbalance circuit to provide control in overrunning loads than to use mechanical counterweights, because the valve setting can be readily adjusted to varying loads (Figure 9.12).

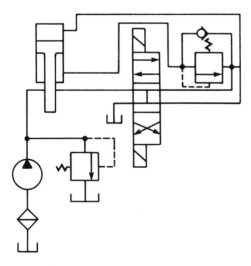

FIGURE 9.12
Counterbalance circuit.

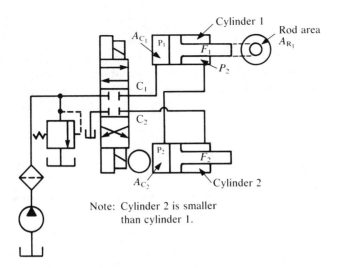

FIGURE 9.13
Synchronizing two cylinders connected in series.

9.10 CIRCUITS FOR SYNCHRONIZING ACTUATORS

The circuit diagram in Figure 9.13 shows a practical, open-loop circuit for synchronizing two cylinders. Cylinders 1 and 2 are connected in series. Hydraulic fluid flows from the pump to the cap end of cylinder 1 through the directional control port C_1. Fluid then flows from the rod end (head) of cylinder 1 to the cap end of cylinder 2. For the two cylinders to be approximately synchronized, the rod-end displacement of cylinder 1 (shown in Figure 9.13) must equal the cap-end displacement of cylinder 2. The pump in this circuit must be capable of delivering pressure equal to that required by the piston of cylinder 1 to overcome the loads on both cylinders, since the cap-end pressure of cylinder 2 acts as a back pressure on cylinder 1. The limitation of this circuit is that over a number of cycles, differences in leakage rates across the pistons will decrease accuracy of the synchronization of the two cylinders.

Since the pressures are equal at the cap end of cylinder 2 and the rod end of cylinder 1, the effects of area, load, and pressure variables can be observed (see Figure 9.13).

Therefore

$$p_1 A_{C_1} - p_2(A_{C_1} - A_{R_1}) = F_1$$
$$p_2 A_{C_2} - p_3(A_{C_1} - A_{R_2}) = F_2$$

By adding both equations and observing that $A_{C_2} = A_{C_1} - A_{R_1}$ and that p_3 is small enough to be equal to zero (due to the return line being open to reservoir), you can reduce these results to

$$p_1 A_{C_1} = F_1 + F_2$$

This equation confirms that in a series open-loop circuit (as shown in Figure 9.13), the pressure will have to be sufficient on A_{C_1} (cap end of cylinder 1) to overcome both forces on the two cylinders. This will be true if we ignore conductor frictional losses and accept that the drain pressure to the reservoir is approximately zero.

9.11 FAIL-SAFE CIRCUITS

Fail-safe circuits are designed to prevent injury to the operator or maintenance personnel or damage to the equipment. Typically, safety circuits prevent an actuator from being actuated in error or prevent a piston from falling. This can be done with a manual reset, push-button directional valve and a pilot-operated check valve as shown in Figure 9.14. This illustration shows that the fail-safe circuit prevents the piston from dropping should a hydraulic conductor rupture or a person mistakenly activate the manual override on the pilot-actuated, directional control valve when the pump is not operating. To lower the piston, pilot pressure from the cap end of the piston must pilot-open the check valve so the oil can return through the directional control valve to the reservoir. This can happen only if the pump is operating and pressure can be regained in the cap end of the piston to remotely open the check valve.

When the push-button valve is actuated to permit pilot-pressure actuation of the directional control valve, the piston will drop. The pilot-operated direction-

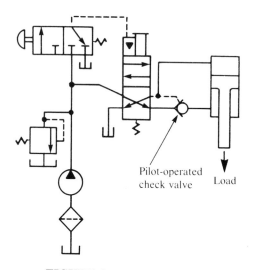

Pilot-operated
check valve   Load

FIGURE 9.14
Fail-safe circuit.

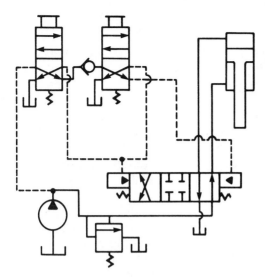

FIGURE 9.15
Two-hand operation safety circuit.

al control valve allows free flow in the opposite direction to retract the piston when the directional control valve returns to its spring offset position. It is important to note that the check valve prevents flow from the rod end except when pilot pressure is sufficient to unseat the ball check valve.

To ensure safe operation, the two-hand operation safety circuit is a very practical application. On presses, it is often necessary to have both of the operator's hands pushing the controls to prevent them from getting in the way of the ram or clamping surfaces. Both manual valves must be operated simultaneously before the piston will extend. Releasing either valve while the cylinder is descending stops the piston. Both push-button manual valves must be released before the piston retracts. Any attempt to tie down the valves to free one's hand prevents raising the piston for a new cycle. Figure 9.15 shows the two-hand operated safety circuit, which must be activated simultaneously to maintain flow pattern.

9.12 LOAD SENSING SYSTEMS

Load sensing control refers to a circuit designed to enhance the sensitivity of control and transmission of power from the prime mover to the load. The advantage of load sensing is that such improved controls conserve energy and reduce or eliminate heat problems in a hydraulic system. The purpose of sensing the load is to determine the precise power needed to attain the desired load movement with the necessary margin of control and safety.

There are two basic methods of sensing load,—with pump controls and with valve controls. Pump controls are probably more popular. However, considerable research and development is being done with valve controls. The load signal may consist of only a load-pressure signal, which may also be combined with a flow signal. As flow increases from a pump output, pressure may also increase rapidly to supply the critical sensing signal to the control valve.

Load sensing pump control systems (Figure 9.16) resemble those in pressure-compensated pumps but with one critical difference. Control pressures for pressure-compensated pumps are sensed inside the pump and reflect the cumulative system pressure variations. However, control pressures for load sensing systems are sensed close to the actual load, reflecting only critical variations in that specific load. In load sensing valve control systems, sensed pressure is used to adjust the setting of a control valve.

Figure 9.16 highlights the pilot control spool valve of a variable-volume pump with a load sensing control. The spool is offset by the mechanical force of a spring equivalent to 150 psi. The spool (spring end) senses the pressure at the outlet of the fixed orifice. When the pressure at the fixed orifice is approximately 150 psi lower than the inlet pressure in the main flow control, the pump will be in equilibrium. It delivers just enough fluid flow so that the pressure loss across the main flow control orifice is at the 150 psi value.

If a reduction in the actuator load occurs, there is an immediate loss in the load-induced pressure that is being fed back to the right-hand spool area. With this reduction in load, there is a loss in equilibrium of force on the pilot spool, causing it to move to the right. Remember, the reduction of load is transmitted as a reduced pressure signal to the spring end. When the pilot spool moves to the right, it vents the control piston to the pump's housing. The small control piston (Figure 9.16) destrokes the pump (reduces its displacement) when the spool vents the larger control spool flow to the pump housing. The remote load sensing causes the pump to respond to the flow demand.

Should the load increase, the induced pressure will increase downstream from the orifice. The increase in load-induced pressure also increases the pressure acting on the pilot control spool and aids the force of the spring, thus moving the pilot spool to the left. When the spool shifts to the left, it loads the larger control piston (Figure 9.16) with system (load) pressure. This signal causes the pump to increase its displacement. The pump will continue to increase its output flow until the resistance to flow creates a pressure at the inlet of the main flow orifice, which is 150 psi higher than the pressure at its outlet.

Note that the pilot relief control in the load sensing control circuit limits the maximum feedback pressure to the pump control. The fixed orifice stabilizes the pump control by limiting the flow in the pilot circuit.

Load sensing allows the variable-displacement pump to respond to the load needs and seek an equilibrium with the load demands. It also can tolerate variations in rpm, provided that maximum flow can be supplied at a minimum drive speed. The pump can decrease its displacement as drive speeds increase by the pilot spool responding to load demands.

We emphasize that a load sensing circuit provides only the flow and pressure required to meet load requirement. Virtually no energy is wasted. Load sens-

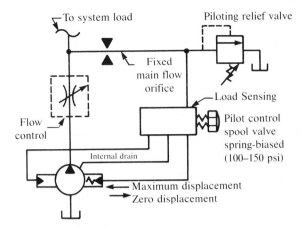

FIGURE 9.16
Typical load-sensing control circuit.

ing circuits deliver lower flow at low pressure, resulting in minimum pump wear and minimum flow and energy losses when the load is constantly sensed.

9.13 SAFETY

Health and safety are very important considerations when working with hydraulic equipment. There are absolutely no compromises with safety whenever hydraulic systems are designed, operated, or maintained. Students, operators, maintenance personnel, design engineers, and technicians working with fluid power components and systems must immediately and constantly correct all unsafe conditions and use safe practices to reduce human error, hazards, and injury to personnel or damage to equipment.

The Occupational Safety and Health Administration (OSHA) of the U.S. Department of Labor provides standards and requirements, and actively enforces these safety standards. OSHA publication 2072, *General Industry Guide for Applying Safety and Health Standards, 29 CFR 1910,* will be helpful to supervisors, managers, employers, trainers, instructors, and students of fluid power. OSHA standards of particular interest to the fluid power industry include Subpart G, Occupational Health and Environmental Control; Subpart N, Material Handling and Storage; and Subpart O, Machinery and Machine Guarding. More recently, fluid power manufacturers must comply with the new Hazard Communications Standard (HCS) of the Occupational Safety and Health Administration. This standard refers to hazardous chemicals. The Industry Guide for Applying Safety and Health Standards 29 CFR 1910 contain standards and requirements, including the following seven categories.

1. **Workplace Standards**   There are basic safety and health standards that apply to the workplace, a building, or other work location. These standards include safety of floors or other working surfaces, access and exit requirements, sanitation, and fire and emergency protection.

2. **Machine and Equipment Standards**   When machines and equipment are present in the workplace, additional elements of risk are present. Standards related to machine guards, operational techniques, special safety devices, inspection and maintenance, and the mounting, anchoring, and grounding of fluid power equipment must be met. Acceptable noise levels are also an important consideration.

3. **Materials Standards**   Materials that are stored, processed, used, or disposed of on the job add to the potential list of hazards. Standards cover such items as toxic fumes, ignitable and/or explosive dusts, and excessive atmospheric contamination. Labeling, material inventory, and employee training for handling hazardous material are also important.

4. **Employee Standards**   Other technical standards become important to reduce potential hazards, such as employee training, personal protective equipment and devices, and medical and first-aid services.

5. **Power Source Standards**   Any power source used in the workplace can also contribute to hazards. Standards are applied to such power sources as electrohydraulic, electrical, pneumatic, hydraulic, and steam-powered systems.

6. **Process Standards**   Specific standards cover a special process or a particular industry. Processes such as abrasive blasting, part dipping, spraying, and welding are hazardous and require protection standards.

7. **Administrative Regulations**   Administrators have numerous responsibilities that they must attend to, including the health and safety of their employees. The employer must display an OSHA poster stating the rights and obligations of employees and employers. Employees are also required to keep safety records related to accidents, illness, and exposure-type occurrences. Fatalities and multiple hospital injury cases including five or more cases have to be reported. An annual summary of injuries and illness must also be posted.

All those in the workplace must be knowledgeable about safe practices and potential sources of hazards. This implies giving instructional attention to prevent personal injury or damage to equipment. It is very important that anyone preparing for a career in the hydraulics industry develop an awareness of personal safety and a positive attitude toward safety and health.

To make students aware of hazards and reduce accidents, the following minimum practices are suggested in a fluid power laboratory

- Wear industrially approved eye protection devices.
- Wear proper clothing. Avoid getting any loose clothing tangled in a rotating or moving part.
- Remember that hydraulic fluid under high pressure if allowed to escape can penetrate your skin easily and cause blood poisoning. Always check that all connections are securely fastened.
- Do not service components or conductors on a control panel or on other equipment while the prime mover is running.
- Before disconnecting any hydraulic lines or pipes, be certain ro relieve all hydraulic pressure in a fluid power system. Quick-disconnect hoses may be uncoupled, since the connectors have check valves to prevent escape of fluid.
- Before working on a circuit equipped with an accumulator, discharge the accumulator.
- When disassembling a fluid power component, do not unload a spring force that may cause parts to scatter.
- Disassemble components in a shallow container to catch any oil escaping from a component and keep the work area oil-free.
- Clean all component parts before reassembling components. Torque all bolts according to manufacturer's specifications.
- Tighten all hose and other conductor connections securely after repairs or replacements are made.
- Test the pressurized components behind a plastic shield or guard to protect you from any hazard.
- Keep your work area and tools clean when servicing any components or assembling circuits. Oil-covered floors are very hazardous.
- Report to the instructor any radical change in operating characteristics such as erratic gauge pressures, unusual component sounds or smells, and leakage from components.
- In case of an injury, seek medical aid at once. Then complete an injury report as soon as possible for an accurate account of what happened.
- Do not operate any hydraulic equipment unless you are familiar with the safety and control functions of the equipment. Your instructor will test you on the use of the equipment or trainer before you are allowed to operate it.
- If you are in doubt about any safety practice, first check with your instructor.
- Any suggestions you have to improve the health or safety of persons working with fluid power systems should be made to the instructor.
- Maintain a positive attitude towards the importance of safety to protect yourself and others from potential hazards at all times.
- Maintain a healthy respect for your equipment.

Summary

Systems design implies a synthesis of an energy transfer system to perform a specific task. It is comprised of interconnected, selected components and conductors to provide required work functions through the transmission, control, and utilization of a pressurized fluid.

A hydraulic system may consist of one or more circuits. In designing a hydraulic system, you begin with the task or load and work through to the various components and ultimately to the prime mover. Unless you are limited to the available prime mover, some compromise may have to be accepted. Testing the system under real operating conditions over many cycles will help confirm its acceptable performance or need for modifications.

The four critical areas of a circuit are: energy input, control, energy output, and auxiliary components. There are many considerations in the selection of alternatives available to perform a given task. Safety, efficiency, reliability, and cost are also determining factors to solve a circuit design problem. It is important to note that the purpose of every hydraulic circuit is to transfer energy (with control) to an output device or actuator for the purpose of doing useful work. The advances in hydraulic technology, particularly in controls (including load sensing controls) may improve even more the efficiency of future hydraulic systems.

Questions and Problems

1. When analyzing or designing a hydraulic circuit, what four critical sections must be considered to produce a safe and workable circuit?
2. What is the principal difference between an open-loop and a closed-loop circuit?
3. How does a constant flow circuit differ from a demand flow circuit regarding positive and instant control?
4. Where would a counterbalance circuit be practical?
5. Explain the purpose of a fail-safe circuit.
6. Draw a schematic circuit of a synchronizing circuit. Explain how synchronization is attained in the circuit.
7. What advantages does a load sensing circuit have over a non-load sensing circuit?

Suggested Learning Activities

1. Compute the horsepower rating of a prime mover for the log splitter circuit (Figure 9.4), if the energy input requirement for the hardwood log splitter was increased to 40,000 lb and the other problem variables remain the same as those listed in the accompanying example problem.
2. Read an article on load sensing and prepare a one-page report explaining the principles involved. Attach a schematic circuit.
3. Hook-up a counterbalance circuit on the trainer and trace the control functions in the circuit.

4. Hook up a synchronizing circuit on the simulator and verify the synchronization over 25 cycles.

Suggested Readings

Esposito, Anthony. *Fluid Power with Applications*. Englewood Cliffs, N.J.: Prentice-Hall, 1980, pp. 256–94.

Frankenfield, Tom. "Using Industrial Hydraulics." 2nd ed. Cleveland, Oh.: *Hydraulics and Pneumatics Magazine,* 1979, pp. 6, 41–46.

Henke, Russell W. "Fluid Power-Systems and Circuits." *Hydraulics and Pneumatics Magazine.* Cleveland, Oh. 1983, pp. 7–43.

Pippenger, John J. *Hydraulic Valves and Controls: Selection and Application*. New York: Dekker, 1984, pp. 224–34.

Sullivan, James A. *Fluid Power: Theory and Applications,* 2nd ed. Reston, Va.: Reston, 1982.

Wolansky, William; Nagohosian, John; and Henke, Russell W. *Fundamentals of Fluid Power*. Boston: Houghton Mifflin, 1977, pp. 214–63.

Yeaple, Frank. *Fluid Power Design Handbook*. New York: Dekker, 1984, pp. 66–77.

CHAPTER 10

Electrical and Electronic Controls

Introduction

Many hydraulic pumps are driven by alternating current (AC) motors that are either single-phase or three-phase systems. Other electrical devices are also used in hydraulic circuits. Some of the more important ones are pushbutton, limit, pressure, temperature, flow, and position switches. There are also coils and contacts, some of which are safety devices and others that generate force to actuate the spools of hydraulic valves.

Electronics also can be used to control the flow of fluid and program the motion of a hydraulic cylinder or motor. The rapid advance of electronic technology in recent years has made possible counters and timers that are essential for process control design. In the field of microelectronics we now have programmable controllers with which data processing is possible and microcomputers that actually perform the function of controlling the hydraulic system.

CPU (Central Processing Unit): An integrated circuit designed to include the necessary circuitry logic functions and everything else needed to provide the signals required of the programmable controller.

Feedback: Monitoring the output of a system so it can be compared with the required output and the actions taken to reduce discrepancies between obtained and required parameters.

Flow switch: A switch actuated by flowing fluid.

Limit Switch: The switch operated when a device has reached its desired maximum, minimum, or any preset value.

Microelectronics: Electronic circuitry integrated into a small element called a *chip*.

Programmable Controller (PC): A device that delivers control signals to a system in a previously designated manner.

Solenoid: A coil of wire that when energized produces an electromagnetic force.

Thermal Switch: A switch operated by a change in temperature.

Time Delay Device: Device used when a time lag is needed between the instant of actuation and the output.

10.1 OPERATION OF PROGRAMMABLE CONTROLLERS

Until recently, the control of many fluid power devices has been performed by electromechanical means. The advantage of programmable controllers is that, in contrast to their analog counterparts, relays, counters, and timers now used in them contain solid state circuitry. An additional benefit is that this type of circuitry is becoming more reliable and less expensive. One further point regarding control systems is that no matter how sophisticated they may be, they are made up of extremely simple elements that are either off or on devices. Other functions the PC can perform are: simple arithmetic operations (exactly like a digital computer), fault locating, raising alarms, operating safety procedures, or long-period system monitoring and regulation.

The three stages of an automated control system are input, logic, and output as shown in Figure 10.1. Most of these modules have a digital property. This means they are devices that are either on or off (+ or −). Examples of input devices are pushbutton contacts, limit switches, and devices by which a variable input can be selected. Note that an input module accepts a + or − signal and converts the signal into a form necessary for use by the PC.

The logic portion of the system includes contacts, coils, and counters, and the output is achieved by driving devices such as motors, heaters, solenoids, and lamps.

10.2 LOGIC LADDER DIAGRAM

A logic ladder consists of a number of parallel linear circuits called rungs, which use the left-hand rail as the input and the right-hand rail as the output. The ladder is provided in a binary format that enables the compiler to translate input and other data into machine language. Some of the symbols used in the ladder logic diagram are shown in Table 10.1.

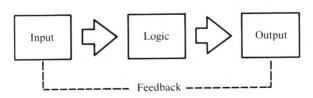

FIGURE 10.1
Schematic sequence of a process.

TABLE 10.1
Symbols used in electrical and electronic circuits.

Device Type	Contact Symbology	
	Normally Open	Normally Closed
Push button		
Limit switch		
Temperature switch		
Flow switch		
Level switch		
Control relay		
Latching relay		
Counter		
Time delay relay. Delay begins when coil is energized		
Delay begins when coil is de-energized		

10.3 APPLICATION TO A HYDRAULIC SYSTEM

One of the simplest fluid power control applications is the actuator of a double-acting hydraulic cylinder through the extend reverse and stop cycle. The fluid power components required to perform this are shown in Figure 10.2a, the sequence diagram in Figure 10.2b, and the logic ladder in Figure 10.2c, showing how the control functions are accomplished. The top rung of the ladder diagram is concerned with the reversing mode, the starting process, a limit switch, and the relay. The lower rung is concerned with the solenoid to activate the valve spool. This is a simple circuit, and frequently the complication of the circuit necessitates the use of many rungs.

The sequence is started by depressing the pushbutton that energizes control relay CR-1. In addition, the coil CR-2, which is placed parallel with the pushbutton, ensures that a circuit is maintained after the start button is released. The control relay CR-1 also closes the contacts CR-3, which energizes the solenoid in the directional control valve Sol-1. As a result, the spool moves to direct oil into the cap end of the cylinder. When the end of the stroke is reached, the NC limit switch 1-LS is opened, which is connected in series with

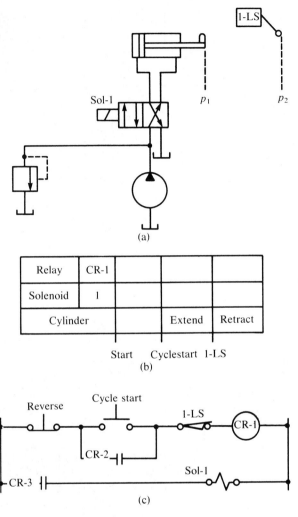

(a)

Relay	CR-1			
Solenoid	1			
Cylinder			Extend	Retract

Start   Cyclestart 1-LS

(b)

(c)

FIGURE 10.2
A system for operating a double-acting cylinder.

CR-2 in the holding circuit. Upon 1-LS opening, the holding circuit is broken, and the relay coil is de-energized. This opens the contacts CR-3, which de-energizes the solenoid. The spring returns the valve spool to its normal position and allows oil to enter the head end of the cylinder. After the cylinder has retracted completely, the reversing is accomplished by activating the NC push-button switch which is connected in series with the start button, the holding contacts CR-2 and CR-3, and the limit switch. Opening this switch breaks the holding circuit coil and, in turn, causes the solenoid circuit contacts to open. This is the same action as that of the circuit switch 1-LS.

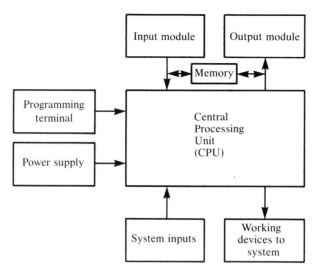

FIGURE 10.3
Basic components of a programmable controller.

10.4 COMPONENTS OF PROGRAMMABLE CONTROLLER SYSTEMS

There are a number of programmable controllers (PCs) provided by many manufacturers, but they do have common parts, as shown in Figure 10.3.

From Figure 10.3 you can see that a typical PC system includes input and output (I/O) modules, memory, a central processing unit (CPU), a means of being programmed, and a power supply. The I/O modules were discussed earlier.

The memory is the device that stores the program or control plan as individual pieces of information called *bits* (or binary digits). The capacity of the memory is important. Capacity is usually measured in kilobytes, where 1 byte is equal to 8 bits.

The CPU scans the control program stored in the memory. It also reads the status of the inputs. The CPU executes commands to the appropriate outputs based on input conditions and what the program is ordering or what performance is required. It also performs such functions as mathematical operations, timing, and counting. In some cases a microprocessor is used as a CPU. The majority of PCs are programmed using a relay logic ladder diagram. The program is entered through the use of a keyboard that prints relay logic symbols.

10.5 PUMP CONTROL

If an electric motor is used as the basic energy input device of a mechanical pumping system, it can be arranged that a combination of almost any pressure up to 5000 psi and any flow up to 200 gpm is possible. The pump, however, has to perform very different tasks in a manufacturing or other sequence. Thus, effective control of the basic quantities is essential. It is also vital to sequence the different operations.

Until recently, control of flow and pressure was accomplished by means of valves, by combinations of pressure feedback from delivery, mechanical devices such as torque limiters, load sensing, or other specialized pump controls. It is, however, becoming clearer that the most convenient way to perform the control functions of a pump or pump-motor combination is by use of microprocessors, which are an extension of the use of electronics principles. This allows an easy interface with the sensing elements, which are usually also electrical. In the material that follows, the basic elements of the control system of a vane-type hydraulic pump are described and explained.

Hydraulic Flow Control

A solenoid, in which the force is proportional to the coil current, moves the sleeve of a stem servo in direct opposition to a mechanical spring. As a result, the pressure in the pump control piston chamber is changed and changes the position of the cam ring by means of an actuator. A feedback link from the cam ring to the stem servo moves the stem to a position of equilibrium within the sleeve. In this manner, the cam ring (displacement of which is proportional to the quantity of oil flowing) is displaced to a position proportional to the solenoid force. The pump flow, in turn, is proportional to cam ring position.

Hydraulic Pressure Control

For this control purpose a solenoid provides force (again proportional to current) that moves a flapper valve in opposition to a spring and hydraulic pressure. In conjunction with another poppet valve, this system controls pressure in the second stage control, again achieving the very desirable property of linearity between the pump pressure and the solenoid current.

10.6 ENERGY EFFICIENCY IN PUMP CONTROL

As a specific example of how a PC can improve the energy efficiency of a pump, consider the three circuits shown in Figure 10.4. They have been designed to provide 2000 psi and 10 gpm for an application where the duty cycle requires 1000 psi and 5 gpm at the cylinder to move a load. What is immediately apparent is the significant improvement achieved by way of reduced energy losses when the PC is used in combination with the pump described in Section 10.4

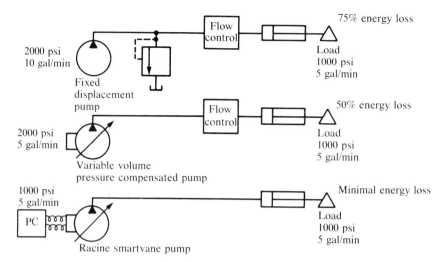

FIGURE 10.4
An energy-efficient pump control system.

10.7 FUTURE PROSPECTS

Athough hydraulic components, seals, fluids, and conductors have undergone considerable design improvements, the most significant advances in hydraulic systems are being developed in the area of controls. Electromechanical controls expanded the diversified use of hydraulics including the servo-valves, but remote sensing was limited as a source of information gathering.

Decision-making and action response were rapidly expanded with the coupling of programmable control logic and production hydraulic control systems. Today's programmable controller makes automatic control systems much more efficient than the earlier electromechanical systems. Essentially the electromechanical relays, counters, timers, and analog devices have been replaced with more reliable solid-state circuitry. The input/output modules provide the critical link between a programmable controller and the hydraulic devices.

Essentially the input-output modules electrically protect the programmable controller from dangerously high voltages. The low DC signal from the PC is applied to an electronic gate, closing a switch to allow AC current to flow in the output devices' circuits. Programmable controllers then use input signals, scan user programs in the logic, solve the ladder rungs, and keep the operation running efficiently and safely.

Thus, the combination of the three technologies—hydraulics, electrical, and electronics—has enabled designers of hydraulics systems to create more efficient, safe, and adaptive circuits. At the same time, while the systems are becoming more complex, integrated, and interdependent, increased instrumentation is necessary to locate, test, and correct malfunctioning components.

The increasingly refined controls applied to hydraulic systems will further augment the expansion of their use. During this century we have taken pride in creating consistent, high-pressure systems that enable potential application of large forces. The focus for the future likely will be directed at controlling energy transmission.

Summary

The elements of electronic and electric mechanisms used in conjunction with hydraulic equipment are given methods of control, and future prospects of the use of electronic devices are outlined. It is concluded that such devices have a promising future. The intent of using electrical and electronic mechanisms is to control the position or velocity of an actuator with greater precision. A programmable controller is a digitally operated electronic device that uses memory for the internal storage of instructions for implementing specific functions such as logic, sequencing, timing, counting, and arithmetic to control various types of machine motions or processes by digital or analog input/output modules.

Questions and Problems

1. What are the three stages of an automated control system?
2. What is meant by "a digital property device"?
3. Describe what happens at the left-hand rail and at the right-hand rail of the logic ladder. What do the rungs represent?
4. Trace the sequence of operations in Figures 10.2a and 10.2b.
5. What are the benefits of controlling a pump output to meet circuit requirements by means of a PC compared to using conventional control valves?

Suggested Learning Activities

1. Write to three different companies who market programmable controllers. List the specifications claimed for each make and construct a table giving a critical comparison between cost complexity and probable effectiveness of operation.

Suggested Readings

Henke, Russell W. "Fluid Power Systems and Circuits." *Hydraulics and Pneumatics Magazine.* Cleveland, Oh., 1983.

Reed, Edward W. and Larman, Ian, S. *Fluid Power with Microprocessor Control: An Introduction.* Englewood Cliffs, N.J.: Prentice-Hall, 1985.

CHAPTER 11 ▬▬▬▬▬▬▬

Pneumatics

Introduction

In the context of fluid power, pneumatics is the study of what happens when a gaseous (compressive) fluid is used instead of hydraulic oil. Air is the medium most often employed due to its plentiful, inexpensive supply on the earth's surface. The principles outlined in this chapter, however, could apply equally to almost any gas. Although used for many decades, pneumatics was extended in the 1900s to many applications through controlled circuitry in industrial processes that required high forces and impact blows. Lightweight and safe pneumatic tools are used in such forms as staplers, pneumatically powered impact wrenches, dentists' drills, riveting guns, and paint sprayers. It is necessary to understand the basic gas laws and pneumatic circuitry to use compressed air effectively for industrial applications. In this chapter you will study the overall properties of air and how air is used in the control and application of pneumatic fluid power.

Absolute Value: Quantity measured from absolute zero of that quantity.

Adiabatic: Any change in which there is no gain or loss of heat.

Boyle's Law: The product of the volume of a gas times its pressure is a constant at a fixed temperature.

Charles' Law: At constant pressure the volume of a fixed quantity of gas varies directly with the pressure.

Degrees Rankine: Absolute value of temperature equal to 460° F.

Gas Constant: A universal constant for a gas that relates pressure, temperature, and volume.

Isothermal Behavior: When the necessary heat for the process is added or subtracted to retain constant temperature.

Ratio of Specific Heats: Ratio of the heat required to raise the temperature of a given mass of gas at constant pressure divided by the heat required at constant volume.

Specific Heat at Constant Pressure and Constant Volume: Heat required to raise one unit mass of gas by one degree in temperature under the conditions stated in the definition.

11.1 PROPERTIES OF AIR

Air is a mixture of gases, the principal constituents being oxygen (21%) and nitrogen (79%), although there are traces of other gases, among them argon and oxides of carbon and nitrogen. Depending upon the atmospheric humidity, air may also contain up to 4 percent water vapor.

Air occupies the entire volume into which it is placed due to the relatively large distance of molecules from one another. This factor is the principal difference between gaseous and liquid fluids. The earth is surrounded by a blanket of air, because air has mass and is attracted to the earth's center. The resulting pressure on the earth's surface at sea level is 14.7 psia. At sea level and a temperature of 32° F air has a density of 0.00238 slug/ft.^{3}, or a specific weight of 0.0766 lb/ft^3.

Bourdon pressure gauges of the type described in Section 2.7 are not sufficiently sensitive to measure atmospheric pressure; for this purpose barometers are used. Barometers typically consist of a column of mercury in an evacuated tube but sometimes a thin diaphragm is used to provide high movement sensitivity to small changes in pressure. The standard atmospheric pressure is 760 mm of mercury, but this figure varies continuously within limits over the earth's entire surface.

Example 11.1 ▶ Show that a standard atmospheric pressure of 760 mm of mercury can be expressed as (a) 29.92 in. of mercury, (b) 34 ft of water, or (c) 14.7 psia.

Solution:

(a) $760 \text{ mm of mercury} = 760 \text{ mm}\left(\dfrac{1 \text{ in.}}{25.4 \text{ mm}}\right) = 29.92 \text{ in. of mercury}$

(b) $29.92 \text{ in. Hg} = \dfrac{29.92 \text{ in. Hg}}{12 \text{ in./ft}}\left(\dfrac{13.6 \text{ ft H}_2\text{O}}{1 \text{ ft Hg}}\right) = 33.9 \text{ ft H}_2\text{O}$

(c) $33.9 \text{ ft H}_2\text{O} = 33.9 \text{ ft}\left(\dfrac{62.4 \text{ lb/ft}^2}{\text{ft}}\right)\left(\dfrac{1 \text{ ft}^2}{144 \text{ in.}^2}\right) = 14.69 \text{ psi}$ ◀

11.2 THE GAS LAWS

When using the gas laws that relate pressure, volume, and temperature to the weight of gas, values of absolute pressure and absolute temperature (mentioned in Chapter 2) must be used.

Conversion for pressure is:

$$psia = psig + patm$$

If we use psi as our unit of pressure then

$$psia = psig + 14.7$$

Also, using the English engineering system of units, repeated from Chapter 2,

$$T_{abs} = T + 460°$$

where T = temperature in degrees Fahrenheit. The units of absolute temperature are degrees Rankine, or °R.

With any change in gas pressure, gas temperature, or gas volume, there will be a change in at least one of the other variables. The behavior of a gas may be examined in relation to the gas laws, which are derived on the basis of a perfect gas. Although a perfect gas does not exist, most diatomic gases, including air, follow the gas laws closely over the range of variables in which we are interested. Considering first a system in which temperature is constant, we may derive the formula for the law devised by Robert Boyle, who stated ". . . provided the temperature of a given mass of air remains constant, its volume varies inversely as the absolute pressure. . . ." If subscripts 1 and 2 signify two separate gas conditions (initial and final conditions), Boyle's law can be written as

$$\frac{p_1}{p_2} = \frac{V_2}{V_1}$$

The relationship however is used more frequently in the form

$$p_2 = p_1\left(\frac{V_1}{V_2}\right) \qquad \text{or} \qquad V_2 = \left(\frac{p_1}{p_2}\right)V_1 \tag{11.1}$$

Example 11.2 ▶ An expansion tank in a heating system is an oil barrel, with an internal diameter of 24 in. and a height of 40 in. When half filled with water the pressure equals 30.0 psig. If the water level falls by 8 in., what will be the pressure of the air in the tank?

Solution:

$$V_1 = 0.5\left(\frac{\pi}{4}\right)24^2(40) = 9047 \text{ in.}^3$$

$$V_2 = 9047 + \left(\frac{\pi}{4}\right)24^2(8) = 12{,}667 \text{ in.}^3$$

$$p_1 = 30 + 14.7 = 44.7 \text{ psia}$$

From Equation 11.1

$$p_2 = p_1 \frac{V_1}{V_2} = 44.7\left(\frac{12{,}667}{9047}\right) = 62.6 \text{ psia}$$

$$\text{final pressure} = 62.6 - 14.7 = 47.9 \text{ psig} \qquad \blacktriangleleft$$

Note that conversions from gauge pressure and degrees Fahrenheit to absolute values must be made to perform the calculations. It is usual, however, to report the final values in the form given in the question.

Now consider a system in which volume is constant. Jacques Charles stated, "... at constant volume, the pressure of a given quantity of air varies in direct proportion to its temperature ..." and "... at constant pressure the volume varies in direct proportion to temperature. ..."

Expressed algebraically

$$\frac{p_1}{p_2} = \frac{T_1}{T_2} \qquad \text{(for constant volume)} \qquad (11.2)$$

and

$$\frac{V_1}{V_2} = \frac{T_1}{T_2} \qquad \text{(for constant pressure)} \qquad (11.3)$$

Example 11.3 ▶ A compressed air receiver has an internal volume of 6000 ft³ and is inflated to a pressure of 135 psig at a temperature of 60° F. Later the temperature is increased by 30° F. Calculate the new pressure of air in the reservoir.

Solution:
Use Equation 11.2

$$p_1 = 135 \text{ psig} = 135 + 14.7 = 149.7 \text{ psia}$$

$$T_1\ 60° F = 460° + 60° = 520° R$$

$$T_2 = 520° R + 30° F = 550° R$$

$$p_2 = p_1\left(\frac{T_2}{T_1}\right) = 149.7\left(\frac{550}{520}\right) = 158.3 \text{ psia}$$

Therefore

$$\text{final pressure} = 158.3 - 14.7 = 143.6 \text{ psig} \qquad \blacktriangleleft$$

11.3 THE CHARACTERISTIC EQUATION FOR A PERFECT GAS

The laws of Boyle and Charles give changes of state of the air when one condition remains constant. Constant conditions are rarely maintained in practice. However, the basic laws can be combined to derive an equation that has universal utility. Thus from Equation 11.1, using p' as the general condition

$$p' = \frac{p_1 V_1}{V_2}$$

and from Equation 11.2

$$p' = \frac{p_2 T_1}{T_2}$$

Equating these two equations for p^1 and transposing terms

$$\frac{p_1 V_1}{T_1} = \frac{p_2 V_2}{T_2} \tag{11.4}$$

Example 11.4 ▶ Fifty cubic feet of air at atmospheric pressure and 68° F are compressed to a pressure of 120 psig, whereupon the gas achieves a temperature of 90° F. Calculate the required volume of the containing vessel.

Solution:
Use Equation 11.4

$$p_1 = 14.7 \text{ psia}$$

$$p_2 = 120 + 14.7 = 134.7 \text{ psia}$$

$$T_1 = 460° + 68° \text{ F} = 528° \text{ R}$$

$$T_2 = 460° + 90° \text{ F} = 550° \text{ R}$$

$$V_1 = 50 \text{ ft.}^3$$

$$V_2 = \left(\frac{p_1}{p_2}\right)\left(\frac{T_2}{T_1}\right) V_1 = \left(\frac{14.7}{134.7}\right)\left(\frac{550}{528}\right) 50 = 5.68 \text{ ft}^3 \qquad ◀$$

Equation 11.4 states that for a given mass of gas, no matter how p, V, and T vary, the ratio pV/T will be a constant quantity. Thus

$$\frac{p_1 V_1}{T_1} = \frac{p_2 V_2}{T_2} = \frac{p_3 V_3}{T_3} = \frac{pV}{T}$$

If we now denote the volume occupied by one pound of gas as V (called the specific volume), then we may write for air

$$\frac{pV}{T} = R \text{ (gas constant)} \qquad (11.5)$$

For one pound mass of air at a pressure of 14.7 psi and 492° F, $R = 53.3$ ft·lb/lb °R.

Other gases have differing values of R, as shown in the following gas table.

Gas	R (ft·lb/lb) °R
Helium (He)	386.0
Hydrogen (H$_2$)	766.4
Nitrogen (N$_2$)	55.2
Oxygen (O$_2$)	48.3

Example 11.5 ▶ A quantity of nitrogen is compressed into a reservoir having a volume of 1800 in.^{3} The final pressure is 150 psig and temperature 100° F. Calculate the mass of the nitrogen in lb.

Solution:

In problems of this type, absolute values are required. In addition it reduces confusion if ft is used as the standard length.

We use Equation 11.5, which refers to 1 lb of gas. Thus, for a mass of gas equal to m lb, $pV = mR$T. From the gas table

$$R = 55.2 \text{ ft·lb/lb °R}$$

$$p = (150 + 14.7) \quad (144) = 23,717 \text{ lb/ft}^2$$

$$V = \frac{1800}{1728} = 1.042 \text{ ft.}^3$$

$$T = 100° + 460° = 560° \text{ R}$$

$$m = \frac{pV}{RT} = \frac{(23,717)\,(1.0421)}{(55.2)\,(560)} = 0.8 \text{ lb} \qquad ◀$$

11.4 COMPRESSION OF AIR

The type of compression considered to this point is that in which time is assumed to have elapsed between the initial state and the final state. This is known as an *isothermal* change and the heat either required or rejected has come from or has been dissipated to the surroundings. If the change in conditions takes place in a short time, as in a compressor, then the heat from generating the compression is retained in the gas. Similarly, when a gas expands suddenly, the heat required to perform this expansion comes from the gas itself and so cooling takes place. This phenomenon is referred to as *adiabatic* behavior.

For an adiabatic process, the relationship between pressure and specific volume is given by

$$p_1 V_1{}^n = p_2 V_2{}^n \qquad \textbf{(11.6)}$$

In Equation 11.6 the quantity n is the ratio for specific heats for air, which has the numerical value 1.4. Since there is always some heat transferred to the surroundings in a real compression, in practice, the exponent to V usually lies between unity and 1.4.

11.5 COMPRESSORS, PIPING, VALVES, ACTUATORS, AND MOTORS

In a similar way to a hydraulic system, the pneumatic system requires, in general, something to generate the pressure, something to convey the air, something to control the flow, and something to utilize the pressure to produce power. Details of the circuitry are given later in this chapter. The details of hardware, however, are excluded due to space limitations. Treatment of such items can be referenced in the bibliography at the end of this chapter.

11.6 PNEUMATIC TRANSMISSION AND DISTRIBUTION SYSTEM

A pneumatic system generates, transmits, and controls the application of power through the use of compressed air within a circuit. A pneumatic system can be illustrated by the three subdivisions that appear in Figure 11.1.

The components of a pneumatic system will include an air compressor and receiver tank to provide the compressed air supply; a filter, regulator, and lubricator (FRL); valves; line pressure gauge; and an actuator (Figure 11.2).

Compressed air is stored in the receiver tank at a predetermined pressure controlled by the safety relief valve. It is transmitted and distributed to various usage devices through interconnected piping, tubing or hose, manifolds, and control valves. The filter is used to remove undesirable contaminants, while the regulator maintains the needed line pressure. The lubricator provides essential lubrication for control valves and actuators to function efficiently.

Clean air extends the life of the compressor and all other components in a pneumatic system. Unlike the situation in hydraulics, compressed air is recir-

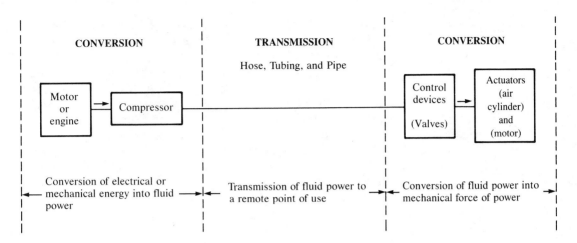

FIGURE. 11.1
Pneumatic system.

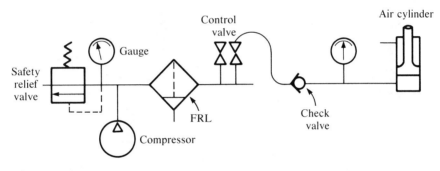

FIGURE 11.2
Circuit components.

culated back to the atmosphere, after performing useful work. A well-designed pneumatic system permits the compressor to supply actuators with the flow rate and pressure needed for the desired operation with a minimum of air leaks at all connections. Although they are not messy like hydraulic leaks, air leaks represent a waste of energy and they contribute to noise levels at the work-place.

11.7 PNEUMATIC POWER CIRCUITS

Pneumatic components are of value when they have been arranged and inter-connected into circuits and systems. A pneumatic circuit can be defined as a group of actuators, controls, and conductors that are interconnected to perform a desired task. A system is a group of circuits intently arranged and interconnected to perform some desired operation.

Presently, air controls are used in a wide variety of industrial applications. The most important reason why these air controls are used is the cost savings generated by the use of pneumatic controls. Miniaturization of components and easy connection of conductors make pneumatic circuits compact and relatively easy to install and maintain.

The basic pneumatic directional control circuit for a single-acting cylinder (Figure 11.3) illustrates the use of a regulated air supply, a directional control valve, and an actuator.

A characteristic of pneumatic circuits is the use of the filter-regulator lubricator unit as the means of the controlled compressed air supply. This unit provides clean air at a regulated pressure and adds enough lubricant to the compressed air to minimize wear of component parts.

Although the two-way directional control valve is able to extend the cylinder, a mechanical spring is necessary to return the piston. Note the absence of a return line from the air exhaust port of the directional control valve, since expanded air is exhausted to the atmosphere.

A time-delay control circuit (Figure 11.4) injects a lag, or a time delay, between the instant at which the actuation signal is applied to the circuit and the moment at which the control valve responds. It is similar to a cushion in hydraulics. In most cases, the time-delay control is adjustable so that the lag can be carried over a given range.

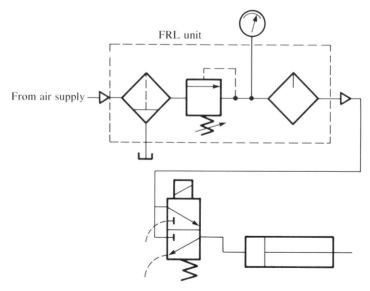

FIGURE 11.3
Basic pneumatic directional control circuit.

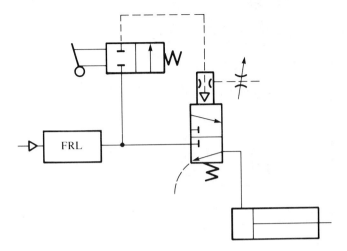

FIGURE 11.4
Time-delay pneumatic circuit.

Flow control pneumatic circuits are more difficult to achieve with varying loads. A typical flow control circuit is illustrated in Figure 11.5. Note that on the extension stroke, the exhaust air is regulated by the metered orifice. On the retraction stroke of the piston, compressed air flows through the check valve route.

Figure 11.6 illustrates a circuit designed to achieve variable speeds of the actuator. A specially shaped cam attached to the piston rod operates the critically located limit switches. These limit switches control the solenoid operators

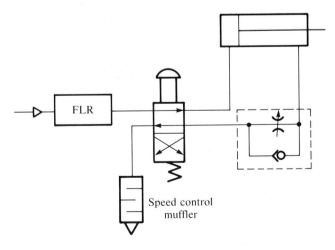

Speed control muffler

FIGURE 11.5
Flow control pneumatic circuit.

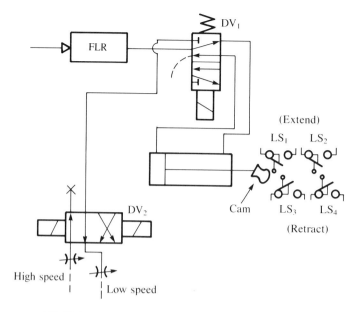

FIGURE 11.6
Multiple-speed pneumatic circuit.

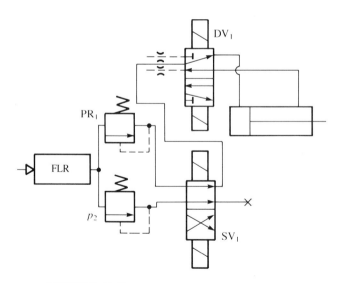

FIGURE 11.7
Dual-pressure circuit.

on the directional control valve DV_2. Each of the ports of DV_2 is connected to a variable-flow control valve. The latter can be set to control the flow of air out of the cylinder and thus the speed of motion of the piston in either direction.

Figure 11.7 illustrates a dual-pressure pneumatic circuit. In this circuit, two pressure regulators are used to supply a single input to the circuit through a selector valve. The circuit can be fed at one of two input pressure levels.

Safety Circuit

Frequently it is advisable to use a two-palm control pneumatic safety circuit such as actuating a press. Both pushbutton-operated pilot valves, A and B, must be actuated simultaneously to extend the piston. Also both pilot valves must be released simultaneously to retract the piston.

Use of Ladder Diagrams

Special purpose elements such as solenoids, relays, and limit switches facilitate interface controls between pneumatics and electrical applications. Ladder diagramming enables a designer to consider inputs, processes, or output condi-

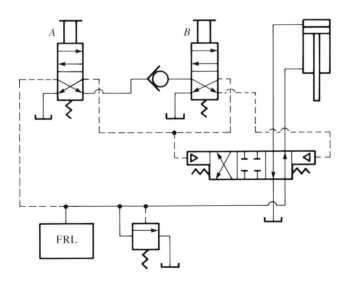

FIGURE 11.8
Pneumatic safety circuit.

tions within a circuit. Examples of pneumatic circuits and ladder diagram circuits are provided in this chapter.

Figure 11.9 shows a circuit that provides continuous reciprocation of air cylinder between two limit switches. When the start button is momentarily pressed, relay coil CR is momentarily energized to shift contactor CR-A, which forms a holding circuit through the stop button and contactor for continuous energization of relay coil CR. Electrical power to the rest of the circuit is also controlled by the relay contactor CR-A. When CR-A is closed, solenoid *A* becomes energized and extends the cylinder, since limit switch 1-LS is held closed. When limit switch 2-LS is held closed by the actuation of the cylinder, solenoid *B* becomes energized and retracts the cylinder. The cylinder continues to travel reciprocally between the two limit switches. When the stop button is depressed, the relay coil becomes de-energized, which removes electrical power from the entire circuit.

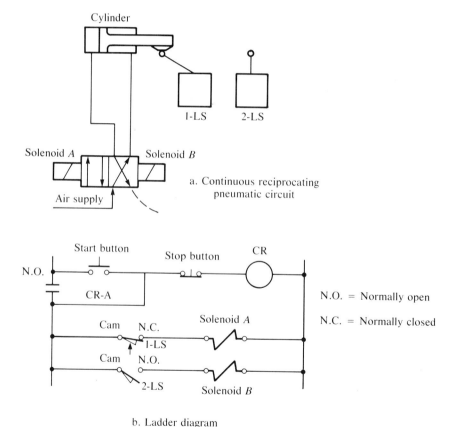

a. Continuous reciprocating pneumatic circuit

b. Ladder diagram

FIGURE 11.9
Continuous reciprocating pneumatic circuit.

Figure 11.10 illustrates a circuit that gives a cycle sequence of two pneumatic cylinders. The cycle sequence initiated by the momentary pressing of the start button is as follows:

1. press start button
2. cylinder 1 extends
3. actuate limit switch 1-LS

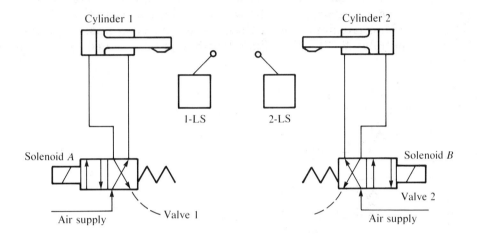

a. Pneumatic sequence circuit

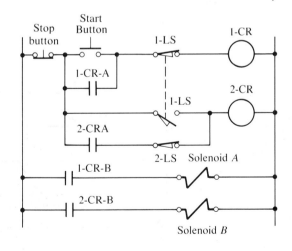

b. Ladder
diagram

FIGURE 11.10
Dual pneumatic cylinder sequence circuit.

4. cylinder 2 extends while cylinder 1 retracts
5. actuate limit switch 2-LS
6. cylinder 2 retracts
7. cycle is completed

When the start button is pressed momentarily, solenoid A is energized to extend cylinder 1. Actuation of limit switch 1-LS by the cylinder 1 de-energizes solenoid A and simultaneously energizes solenoid B, because limit switch 1-LS is a double-pole, single-throw type. Actuation of limit switch 1-LS breaks the holding circuit for relay 1-CR and simultaneously closes the holding circuit for the relay 2-CR. When limit switch 2-LS is actuated by the cylinder 2, the

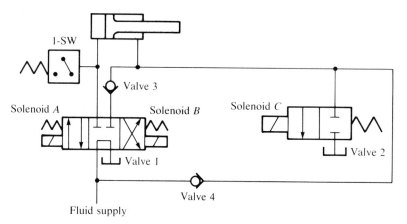

a. Hydraulic circuit

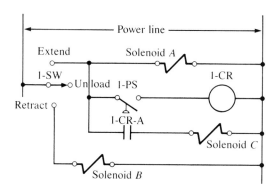

b. Electrical ladder diagram

FIGURE 11.11
Regenerative circuit.

holding circuit for relay 2-CR is broken, which de-energizes solenoid B to shift valve 2 back to its spring offset mode, thus causing the retraction of the cylinder 2. The stop button is used to break the source of electrical power in the circuit and instantly retracts both cylinders.

Figure 11.11 illustrates a regenerative circuit controlled with two solenoid valves, a pressure switch, and two check valves. The circuit operation is as follows: when switch 1-SW is manually turned to the extend position, solenoid A is energized, which causes the cylinder to extend. Oil from the rod end flows through check valve 4 to join the incoming oil from the pump to provide a rapid cylinder extension in the regenerative condition. When oil pressure builds up due to loading on the cylinder to actuate a normally-open pressure switch 1-PS, relay 1-CR and solenoid C become energized. Rod-end oil is vented directly back to the oil tank through valve 2. As a result, the cylinder extends slowly as it drives a high resistive load. Relay contact 1-CR-A provides a holding circuit for relay coil 1-CR and continues to energize solenoid C during the slow extension of the cylinder, which prevents any fluttering of the pressure switch. It can be seen that the pressure switch is by-passed during the cylinder slow extension portion of the cycle.

When switch 1-SW is turned to the retract position, solenoid B becomes energized, while the relay coil and solenoid C become de-energized. As a result, the cylinder retracts in a normal manner to its fully retracted position. When the switch is turned to the unload position, all the solenoids and relay coils are de-energized. This places valve 1 in its spring-centered position to unload the pump.

Ladder diagrams provide a circuit designer with a practical means to plan efficient control function systems. It is necessary to consider all parts of a circuit for the entire operational cycle and repeated cycles, as well as to consider the aspect of safety should electrical power failure occur. Ladder diagrams can be used for pneumatic as well as hydraulic circuits.

Summary

This chapter introduced you to the gas laws, which enable you to understand that gases behave differently from liquids in a fluid power system. Pneumatic circuits withstand lesser pressure ratings than do hydraulic circuit components.

The principles of circuit design do not vary much, not withstanding that air is more compressible than liquids. Essentially, a circuit designer has to start with the load or work function to be achieved and determine the most efficient actuators, controls, and input devices required.

Electrical and electronic devices can be interfaced with pneumatic and hydraulic components to provide efficient and safe circuits.

Ladder diagrams provide the circuit designer with a practical means to examine input, process, and output functions to plan quickly the circuit layout design for a particular pneumatic application.

Questions and Problems

1. If the average air pressure over the earth's surface is 14.69 psia, calculate the mass of air surrounding the earth given that it is 7950 miles in diameter. (Note: the surface area of a sphere of radius r is $4\pi r^2$).

2. An air receiver has an end bladder that facilitates changes in volume in order to maintain constant pressure. When the temperature of the air is 50° F, its volume is 5000 ft.^{3} If no air is added or removed and its temperature is increased to 110° F, calculate the final volume.

3. A pneumatic accumulator has a volume of 2500 in.^{3} when under pressure of 1400 psig and at a temperature of 100° F. The volume is reduced to 2200 in.^{3}, while the temperature changes to 300° F. What is the final gauge reading?

4. Given: 4 lb of helium at a pressure of 150 psig and a temperature of 90° F. What is the volume of the helium under these conditions?

Suggested Learning Activities

Visit a local industrial plant where pneumatic power is used either for control or for performance of an engineering task. Diagram the layout of the pneumatic system and note carefully the physical locations of the compressor, FRL unit, and the controls in relation to the actuators.

Suggested Readings

Esposito, Anthony. *Fluid Power with Applications.* Englewood Cliffs, N.J.: Prentice-Hall, 1980, 295–338.

Henke, Russell W. "Fluid Power Systems and Circuits." *Hydraulics and Pneumatics Magazine.* Cleveland, Oh., 1983, pp. 103–10.

————. *Industrial Pneumatic Technology.* Bulletin 0275-Bl. Cleveland, Ohio: Parker Fluid Power, 1980, pp. 1–34.

Johnson, Olaf F. *Fluid Power: Pneumatics.* Melbourne, Fla.: Krieger, 1985, pp. 1–65.

Lansky, A.L. and Schrader, L.F. *Industrial Pneumatic Control.* New York: Dekker, 1986, pp. 63–91.

McCord, Bruce E. *Designing Pneumatic Control Circuits.* New York: Dekker, 1983, pp. 79–110.

Newton, Donald G. *Fluid Power for Technicians.* Englewood Cliffs, N.J.: Prentice-Hall, 1971, pp. 58–70.

Randall, Ralph D. *Fundamentals of Pneumatic Control,* St. Regis Paper Company: 1969, pp. 7g, 1a, and 8, 1–9.

Sullivan, James A. *Fluid Power Theory and Applications,* 2nd ed. Reston, Va.: Reston, 1982, pp. 335–78.

Wolansky, William; Nagohosian, John; and Henke, Russell W. *Fundamentals of Fluid Power.* Boston: Houghton Mifflin, 1977, pp. 249–59.

APPENDIX A

Properties of Some Common Commercial Hydraulic Fluids and of Water

	MIL-H-5606B	Aircraft Phosphate Ester	Aircraft Silicate Ester	Water
Specific gravity (sg), 60° F	0.848	1.07	0.93	1.0
γ lb/ft^3	52.9	66.8	58.0	62.4
ρ slug/ft^3	1.64	2.07	1.80	1.93
Viscosity, cS °F				
−65	2130	2300	2357	—
−40	500	480	600	—
0	100	90	150	—
100	14.3	11.5	24.3	0.69
210	5.1	3.9	8.1	0.30
400	1.9	—	2.6	—
Specific Heat Btu/lb, °F	0.5	0.38	0.44	1.0
Flash point, °F	225	360	395	none

Factors in the selection of fluids

Fluid	Advantages	Disadvantages
Air	Available universally Relatively clean Easily harnessed Can be exhausted to atmosphere Clean leakage	Expensive Limited to low pressure Subject to moisture

PROPERTIES OF SOME COMMON COMMERCIAL HYDRAULIC FLUIDS AND OF WATER

Dry nitrogen	Clean Safe Inert	Expensive Not readily available Sponginess (sometimes)
Petroleum oils	Excellent lubricity Reasonable cost Noncorrosive	Tendency to oxidize rapidly No fire resistance
Phosphate esters	Excellent fire resistance Good lubricity Noncorrosive	Not compatible with some pipe compounds and some seals Fairly expensive No rust protection
Water	Inexpensive Readily available Fire-resistant	No lubricity Corrosion-inducing Temperature limitations
Water-glycol	Good fire resistance Inexpensive Compatible with most pipe compounds and seals	Not good for radial-piston pumps Poor corrosion resistance May cause excessive pump wear at high pressures
Water-oil emulsion	Good fire resistance Inexpensive Compatible with most seals	Sometimes difficult to maintain Low lubricity

APPENDIX B

Frequently Used Nomenclature

Symbol	Definition	Units
ROMAN ALPHABETIC SYMBOLS		
a	Acceleration	in./s^2, ft/s^2, cm/s^2, m/s^2
A	Area	in.^{2}, ft^2, cm^2, m^2
BP	Burst pressure	psi (lb/in.^{2})
C_d	Discharge Coefficient	dimensionless
d	Diameter	in., ft, cm, m
D	Diameter	in., ft, cm, m
D_i	Inside diameter	in., ft, cm, m
D_o	Outside diameter	in., ft, cm, m
e	Eccentricity	in., ft, cm, m
E	Energy	lb·in., lb·ft, N·cm, N·m
F	Force	lb, N
f	Friction factor	dimensionless
g	Acceleration of gravity	in./s^2, ft/s^2, cm/s^2, m/s^2
H	Head	in., ft, cm, m
HP	Horsepower	hp
H_L	Head loss	in., ft, cm, m
H_m	Motor head	in., ft, cm, m
H_p	Pump head	in., ft, cm, m
ID	Inside diameter	in., ft, cm, m
J	Joule	N·m
K	K factor spring stiffness	lb/in., N/m
KE	Kinetic energy	lb·in., lb·ft, N·cm, N·m
L	Length	in., ft, cm, m
L_e	Equivalent length	in., ft, cm, m
m	Mass	slugs, kg
n	Rotational (speed)	rpm
N	Force in newtons	kg·m/s^2

235

OD	Outside diameter	in., ft, cm, m, mm
P	Power	ft·lb/s, J/s
p	Pressure	lb/in.^{2}, lb/ft^2, N/cm^2, N/m^2
Pa	Pascal	N/m^2
Q	Volumetric flow rate	in.^{3}/s, ft^3/s, cm^3/s, m^3/s, gpm
Q_A	Actual flow rate	in.^{3}/s, ft^3/s, cm^3/s, m^3/s, gpm
Q_T	Theoretical flow rate	in.^{3}/s, ft^3/s, cm^3/s, m^3/s, gpm
r	Moment arm	in., ft, cm, m
Re	Reynolds number	dimensionless
S	Distance	in., ft, cm, m, mm
SF	Safety factor	dimensionless
sg	Specific gravity	dimensionless
t	Wall thickness	in., cm, mm
t	Time	s
T	Temperature	°F, °R
T	Torque	lb·in., lb·ft, N·cm, N·m
T_A	Actual torque	lb·in., lb·ft, N·cm, N·m
T_T	Theoretical torque	lb·in., lb·ft, N·cm, N·m
v	Velocity	in./s, ft/s, cm/s, m/s
V	Volume	in.^{3}, ft^3, cm^3, m^3
V_D	Volumetric displacement	in.^{3}, ft^3, cm^3, m^3
VI	Viscosity index	dimensionless
w	Weight flow rate	lb /s, lb/min
W	Weight	lb, N
W	Work	lb·in., lb·ft, N·cm, N·m
W	Watt	J/s
WP	Working pressure	lb/in.^{2} (psi)
y	Oil film thickness	ft
Z	Elevation	in., ft, cm, m
Δp	Pressure drop	lb/in.^{2}, lb/ft^2, N/cm^2, N/m^2

GREEK ALPHABETIC SYMBOLS

Symbol	Definition	Units
γ	Specific weight	lb/in.^{3}, lb/ft^3
μ	Dynamic (or absolute) viscosity	lb·s/ft^2, gm/cm·s
ν	Kinematic viscosity	ft^2/s, cm^2/s
π	Ratio of circle perimeter to its diameter (numerical value = 3.1416)	dimensionless
ρ	Mass density (mass per unit volume)	slug/ft^3

APPENDIX C

Important Fluid Power Facts and Formulas

Theory	Equation Number
Bernoulli Equation	**(3.3)**
$$\frac{p}{\gamma} + \frac{v^2}{2g} + h = \text{constant}$$	
Continuity Equation	**(3.1)**
$$Q = Av = \text{constant}$$	
Reynolds Number	**(3.12)**
$$\text{Re} = \rho\frac{vD}{\mu}$$	
Darcy-Weisbach Equation	**(3.13)**
$$H_F = f\left(\frac{L}{D}\right)\frac{v^2}{2g}$$	
Venturi/Orifice Equation	**(3.10)**
$$Q = C_d A \sqrt{\frac{2(\Delta p)}{\rho\,[1 - (A_2/A_1)^2]}}$$	

Fluids

Oil is the most commonly used hydraulic fluid because it serves as a lubricant for hydraulic components and is virtually incompressible (0.004% at 1000 psi). It has a specific weight of approximately 58 lb/ft^3.

Pressure at the bottom of a 1 ft column of oil is approximately 0.4 psi (58/144). To find the approximate pressure at the bottom of an oil column, multiply the height in feet by 0.4.

Area

area of a circle $= \pi r^2 = \dfrac{\pi D^2}{4} = 0.7854 D^2$

Volume

1 U.S. gallon = 231 in.^{3}

1 ft^3 = 1728 in.^{3}

1 ft^3 = 7.48 gal

Cylinders

area $= D^2 \times 0.7854$

force output (lb) = psi $\times$ effective area (in.^{2}) = pA

The speed of a cylinder is dependent on its effective piston area and the rate of fluid flow.

$$\text{piston speed (in./min)} = \frac{\text{pump output (in.}^3\text{/min)}}{\text{effective area of piston (in.}^2)}$$

$$\text{or} \quad \frac{231 \times \text{gpm}}{\text{area}}$$

displacement of a hydraulic cylinder (in.^{3}) = effective area (in.^{2}) $\times$ stroke (in.)

$$\text{time to extend piston (min)} = \frac{\text{cylinder displacement (in.}^3)}{\text{pump output (in.}^3\text{/min)}}$$

Conductors

Flow velocity through a pipe varies inversely as the square of the inside diameter. Doubling the size of the D_i will increase the area four times.

$$\text{velocity (ft/s)} = \frac{231 \times \text{gpm}}{12 \times 60 \times \text{area}} \quad \text{or} \quad \frac{0.3208 \times \text{gpm}}{\text{area}}$$

rate of flow (Q) = area $\times$ velocity

To find the area of a pipe needed to handle a given flow, use this formula:

$$\text{area} = \frac{\text{gpm} \times 0.3208}{\text{velocity (ft/s)}}$$

Power

1 hp = 550 ft·lb/s

7.48 gal = 1 ft^3

Then

$$1 \text{ gpm} = \frac{1}{7.48 \times 60} \text{ ft}^3/\text{s}$$

$$1 \text{ psi} = 1 \text{ lb/in.}^2 \times \frac{144 \text{ in.}^2}{\text{ft}^2} = 144 \text{ lb/ft}^2$$

Therefore, the energy required to pump 1 gal at 1 psi in 1 s is as follows:

$$\text{energy} = \frac{1}{7.48 \times 60} \times 144 = 0.321 \text{ ft·lb/s}$$

Converting the above quantity of energy to horsepower:

$$\text{energy} = \frac{0.3208}{550} = 0.000583 \text{ hp}$$

$$HP = \text{gpm} \times \text{psi} \times 0.000583$$

Horsepower input to pump:

$$HP_{\text{in}} = \frac{\text{gpm} \times \text{psi} \times 0.000583}{e_{\text{o}}}$$

e_{o} presents the overall efficiency of a unit.

$$\frac{HP_{\text{out}}}{HP_{\text{in}}}$$

Motors

$$\text{torque (lb·in.)} = \frac{\text{in.}^3/\text{rev} \times \text{psi}}{2\pi}$$

$$\begin{array}{l}\text{theoretical} \\ \text{motor} \\ \text{speed}\end{array} = \frac{\text{theoretical displacement of pump (in.}^3/\text{rev)} \times \text{pump speed (rpm)}}{\text{theoretical displacement of motor (in.}^3/\text{rev)}}$$

$$= \text{rev/min of motor}$$

The actual speed will vary with the volumetric efficiency of both the pump and the motor.

APPENDIX D

Conversion of English Gravitational Unit System to the SI System

Quantity	English Unit	SI Unit	Metric Symbol	Equivalent Unit
Length	1 foot (ft)	0.3048 meter	m	—
Mass	1 slug	14.59 kilograms	kg	—
Time	1 second (sec)	1.0 second	s	—
Force	1 pound (lb)	4.448 newtons	N	kg·m/s
Pressure	1 lb/in^2 (psi)	0.3325 pascals	Pa	N/m^2 or kg/(m·s^2)
Temperature	Fahrenheit	°R/1.80 = K	°K	—
Energy	1 ft·lb	1.356 joule	J	kg·m^2/s or m·N
Power	1 ft·lb/s	1.356 watt	W	J/s or N·m/s

$$T_{°R} = T_{°F} + 459.67 = 1.80(T_{°C}) + 491.67$$

SUPPLEMENTARY CONVERSION UNITS

English Unit	SI Unit	SI Symbol	Equivalent Unit
Length			
1 inch (in.)	2.540000 centimeters	cm	25.400000 mm
1 foot (ft)	0.304800 meter	m	30.4800 cm
1 yard (yd)	0.914400 meter	m	91.4400 cm
1 mile (mi)	1.609344 kilometer	km	1609.344 m
Area			
1 in.^{2}	6.451600 square centimeters	cm^2	
1 ft^2	0.092903 square meter	m^2	
Volume			
1 in.^{3}	16.387064 cubic centimeters	cm^3	
1 ft^3	0.028317 cubic meter	m^3	
Velocity			
1 ft/s	30.480 centimeters/second	cm/s	
1 ft/s	0.304800 meter/second	m/s (preferred)	
1 mile/hour	1.609344 kilometer/hour	km/h	

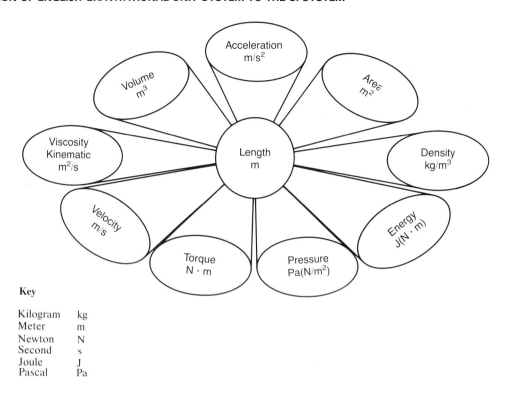

Key

Kilogram	kg
Meter	m
Newton	N
Second	s
Joule	J
Pascal	Pa

English Unit	SI Unit	SI Symbol
Volume flow rate		
1 in.^{3}/s	16.3871 cubic centimeters/second	cm^3/s
1 in.^{3}/min	983.224 cubic centimeters/second	cm^3/s
1 ft^3/sec	28.31685 liters/second	l/s
1 ft^3/min	1699.011 liters/second	l/s
1 U.S. gal/s	3.785412 liters/second	l/s
1 U.S. gal/min	227.1247 liters/second	l/s
Mass		
1 slug	14.606 kilogram	kg
1 lb	0.453592 kilogram	kg
Pressure		
Standard atmosphere (14.7 psi)	101.352 kilopascals	kPa
1 inch of water (at 4° C or 39.2° F)	249.082 pascals	Pa

English Unit	SI Unit	SI Symbol
1 lb force per square inch (psi)	6.894757 kilopascals	kPa
Energy 1 foot·pound (ft·lb)	1.355818 joule	$J = N \cdot m$
Force 1 lb·(lb$_f$)	4.448222 newton	N
1000 lb	4.448222 kilonewton	kN
Power 1 ft·lb/s	1.355818 watts	W
1 ft·lb/min	0.022597 watt	W
1 U.S. hp (550 ft·lb/s)	745.700 watts	$W = 1\ J/s$ or $= 1\ N \cdot m/s$
Kilowatts	$\dfrac{\text{torque(newton·meters)} \times \text{rpm}}{0.000955}$	kW
Torque 1 lb force·inch (lb·in.)	0.112985 newton·meter	$N \cdot m$
1 lb force·foot (lb·ft)	1.355818 newton·meter	$N \cdot m$
Viscosity pound force s/foot (lb·s/ft)	47.88026 pascal·seconds	$Pa \cdot s$
square foot/second (kinematic); ft^2/s	0.092903 square meter/second	m^2/s

PREFIXES FOR DECIMAL MULTIPLES AND SUBMULTIPLES

The prefix scheme used in the SI system is the scientific notation system. This system uses a base-10 function. There are four prefixes for multiples of the base units (Table D.1) and seven prefixes for the submultiples of the base units (Table D.2). Note that preferred units have an index to the base-10 that is divisible by three.

TABLE D.1
Multiples

Prefix	Symbol Added to Base Unit	Value	Power of 10
kilo	k	1 000 units	10^3
mega	M	1 000 000 units	10^6
giga	G	1 000 000 000 units	10^9
tera	T	1 000 000 000 000 units	10^{12}

TABLE D.2
Submultiples

Prefix	Symbol Added to Base Unit	Value	Power of 10
centi	c	0.01 unit	10^{-2}
milli	m	0.001 unit	10^{-3}
micro	μ	0.000 001 unit	10^{-6}
nano	n	0.000 000 001 unit	10^{-9}
pico	p	0.000 000 000 001 unit	10^{-12}
femto	f	0.000 000 000 000 001 unit	10^{-15}
atto	a	0.000 000 000 000 000 001 unit	10^{-18}

APPENDIX E

Symbols Used for Fluid Power (ANSI)

Lines and Line Functions

———	Line, working	—→—	Direction of flow, hydraulic
− − − − −	Line, pilot	⊥	Line to reservoir above fluid level below fluid level
- - - - -	Line, drain		
•	Connector		
⌣	Line, flexible	⊣	Line to vented manifold
┼	Line, joining	✕	Plug or plugged connection
┿	Line, passing	≍	Restriction, fixed

Pumps	
	Pump, single fixed displacement
	Pump, single variable displacement
Motor and Cylinders	
	Motor, rotary fixed displacement
	Motor, rotary variable displacement
	Motor, oscillating
	Cylinder, single acting
	Cylinder, double acting
	Cylinder, differential rod
	Cylinder, double end rod
	Cylinder, cushions both ends
Basic Valve Symbols	
	Check valve
	Manual shut off valve
	Basic valve envelope
	Valve, single flow path, normally closed

	Valve, single flow path, normally open
	Valve, maximum pressure (relief)
	Basic valve symbol multiple flow paths
	Flow paths blocked in center position
	Multiple flow paths (arrow shows flow direction)
Miscellaneous Units	
	Direction of rotation (arrow in front of shaft)
	Component enclosure
	Reservoir, vented
	Reservoir, pressurized
	Pressure gauge
	Temperature gauge
	Flow meter (flow rate)
	Electric motor
	Accumulator, spring loaded
	Accumulator, gas charged
	Filter or strainer
	Heater
	Cooler
	Temperature controller
	Intensifier
	Pressure switch

Valve Examples	
	Unloading valve, internal drain, remotely operated
	Deceleration valve, normally open
	Sequence valve, directly operated, externally drained
	Pressure reducing valve
	Counter balance valve with integral check
	Temperature and pressure compensated flow control with integral check
	Directional valve, two position, three connection
	Directional valve, three position, four connection
	Valve, Infinite positioning (indicated by horizontal bars)

Methods of Operation	
	Pressure compensator
	Detent
	Manual
	Mechanical
	Pedal or treadle
	Push button
	Lever
	Pilot pressure
	Solenoid
	Solenoid controlled, pilot pressure operated
	Spring
	Servo

APPENDIX F

Sizes of Steel Pipe

Nominal Pipe Size (in.)	Outside Diameter (in.)	Inside Diameter (in.)	Wall Thickness (in.)	Internal Area (in.²)
		Schedule 40		
⅛	0.405	0.269	0.068	0.0568
¼	0.540	0.364	0.088	0.1041
⅜	0.675	0.493	0.091	0.1910
½	0.840	0.622	0.109	0.304
¾	1.050	0.824	0.113	0.533
1	1.315	1.049	0.133	0.864
1¼	1.660	1.380	0.140	1.496
1½	1.900	1.610	0.145	2.036
2	2.375	2.067	0.154	3.36
2½	2.875	2.469	0.203	4.79
3	3.500	3.068	0.216	7.39
3½	4.000	3.548	0.226	9.89
4	4.500	4.026	0.237	12.73
5	5.563	5.047	0.258	20.01
6	6.625	6.065	0.280	28.89
8	8.625	7.981	0.322	50.0
10	10.750	10.020	0.365	78.9
12	12.750	11.938	0.406	111.9

SIZES OF STEEL PIPE

Nominal Pipe Size (in.)	Outside Diameter (in.)	Inside Diameter (in.)	Wall Thickness (in.)	Internal Area (in.²)
		Schedule 80		
⅛	0.405	0.215	0.095	0.0364
¼	0.540	0.302	0.119	0.0716
⅜	0.675	0.423	0.126	0.1405
½	0.840	0.546	0.147	0.2340
¾	1.050	0.742	0.154	0.432
1	1.315	0.957	0.179	0.719
1¼	1.660	1.278	0.191	1.283
1½	1.900	1.500	0.200	1.767
2	2.375	1.939	0.128	2.953
2½	2.875	2.323	0.276	4.24
3	3.500	2.900	0.300	6.61
3½	4.000	3.364	0.318	8.89
4	4.500	3.826	0.337	11.50
5	5.563	4.813	0.375	18.19
6	6.625	5.761	0.432	26.07
8	8.625	7.625	0.500	45.7
10	10.750	9.564	0.593	71.8
12	12.750	11.376	0.687	101.6

ANSWERS

Answers to Even Number Study Questions and Problems

Chapter 1

2. The minimum list of components to construct a hydraulic circuit are: prime mover, pump and reservoir, relief valve, directional valve, an actuator, and conductors.

4. Sketch to include the following: prime mover, power steering pump, control valve, hose connections, actuator, and reservoir.

6. Key points to include: development of the self-contained pump, reservoir, and prime mover unit; the impact of petroleum fluid to replace water; improved seals; improved machining; increasing pressure systems; interfacing of electrical/hydraulic/mechanical and ultimately electronics and programmable controllers; miniaturization of hydraulic components; and extremely high pressure systems and control of large forces.

Chapter 2

2. 4800 lb·ft

4. 6.44 ft/s^2

6. 0.729 lb·ft

8. (a) 46.7 lb·ft
   (b) 24.43 ft·lb
   (c) 58.6 ft·lb/s

10. 800 lb, 900 ft·lb

12. (a) 17.3 slugs, (b) 1.725 slug/ft^3, (c) 55.56 lb/ft^3

14. col 2 2.72×10^{-3}
    col 3 1.89×10^{-5}
    col 4   0.13
    col 5   1.47
    col 6 1.58×10^{-3}
    col 7 1.47×10^{-4}

16. 53.2 in./s

Chapter 3

2. (a) 470; (b) 66.2
4. (a) 14.3 ft/s
   (b) 2.5 ft/s
6. 1504 psi
8. 0.63
10. (a) 361 (laminar)
    (b) 8335 (turbulent)
12. 31.9 ft of oil, 13.6 psi
14. 0.419 slugs/s, 92.5 psi
16. An analytical problem. Substitution suggested in text.

Chapter 4

2. Many hydraulic systems problems are caused by contaminated fluid. Grit, scales from pipe, chips from machined parts, tape from improper application to threaded pipe, and water can all be harmful to precision-machined components. For these reasons, fluid should be kept clean and strained while being replaced into a cleansed system.

4. The three types of conductors are: rigid pipe, tubing, and flexible hose.

6. For a pipe to conduct 5 gallons of fluid per minute at a velocity of 20 feet per second, the diameter would have to be

$$A = \frac{5 \times 0.3208}{20}$$

$$A = \frac{\pi D^2}{4}$$

$$A = 0.7854 D^2$$

$$D^2 = \frac{0.0802}{0.7854}$$

$$D^2 = 0.1021$$

$$D = 0.3195 \text{ in.}$$

8. Flexible hosing would be used where there is any movement or excessive vibration between two movable segments of a machine. Examples might include the extension of a long stroke cylinder such as on a front-end loader or a swivel where one component twists in relation to the other, such as a retractable landing gear on a plane.

10. Four considerations in selecting a seal are: hardness, resilience, abrasion resistance, and load-bearing capacity.

Chapter 5

2. Viscosity is a measure of the internal friction or the resistance of a fluid to flow, usually expressed in units of centipoise. Viscosity index (VI) indicates the change in viscosity with temperature. A low viscosity index indicates a small change in viscosity for given rise in temperature.

4. Three types of fire-resistant hydraulic fluids are: water-oil emulsions, phosphate esters, and chlorinated hydrocarbons.

6. A strainer is a device generally designed with a fine wire mesh screen to remove coarse particles from a fluid. A filter is a device whose primary function is the removal and retention by a porous medium of insoluble contaminants.

8. An absorbent filter contains an inactive medium such as cotton, paper, and other fibers to trap the contaminants by the walls of the medium. An adsorbent medium is active in that the element materials adhere to the surface of certain suspended contaminants. Adsorption is a surface phenomenon. Charcoal or chemically treated paper can adsorb unwanted contaminants from a fluid.

10. If the pump is 92 percent efficient, it would require a prime mover of the following horsepower

$$\text{power} = \text{pressure (psi)} \times \text{flow rate } (Q)$$

$$= 1323 \text{ psi} \times 65 \text{ gpm}$$

$$= \frac{1323 \times (12)^2 \text{ lb}}{\text{ft}^2} \times \frac{65 \ (231) \text{ in}^3}{\text{min}}$$

$$= \frac{1323 \times (12)^2 \text{ lb}}{\text{ft}^2} \times \frac{65 \times 231 \ (1/12)^3 \text{ ft}^3}{60 \text{ s}}$$

$$= 27{,}590 \text{ ft} \cdot \text{lb/s}$$

$$HP \quad = \frac{27{,}590}{550} = 50.16$$

or

$$HP \quad = \text{gpm} \times \text{psi} \times 0.000583$$

$$= 65 \times 1323 \times 0.000583$$

$$= 50.14$$

The input horsepower would have to be greater since the pump is 92 percent efficient.

$$HP = \frac{50.16}{0.92} = 54.52$$

Chapter 6

2. A volumetric void is created by changing the internal contours. This gives rise to a partial vacuum, causing fluid to be "sucked in." Judicious use of the tables and diagrams in the text and the Suggested Readings is then required.

4. 1.68 in.^{3}/rev

6. 1.25 in.

8. 88.3%

10. 95.3%

Chapter 7

2. A simple relief valve is a direct type of relief valve wherein the system pressure opposes a poppet or piston that is held on its seat by an adjustable spring. The spring tension is set to limit the maximum pressure that can be maintained within the system. The compound relief valve is a two-stage design: the main spool is balanced by system pressure and the spring maintains the spool seated. The pilot valve actually controls the opening of the main spool. The spring is adjusted on the pilot valve rather than on the main valve.

4. An unloading valve is designed primarily to unload a pump back to the reservoir at minimum resistance. Return flow over a relief valve encounters more resistance and is a waste of horsepower.

6. The horsepower loss across a relief valve for a fixed 20 gpm pump and 50 psi is

$$
\begin{aligned}
HP &= \text{gpm} \times \text{psi} \times 0.000583 \\
&= 20 \times 50 \times 0.000583 \\
&= 0.583
\end{aligned}
$$

8. *Ways* in a control valve indicates the number of paths in which a fluid flows. A one-way check valve means fluid flows in one direction only. It has to have an inlet and an outlet port. There is only one flow path. Position refers to the control that a spool or other device has to select given flow paths. A three-position directional control valve alters the flow pattern each time the position of the control device is moved.

10. Symbols for the valves include:
    (a) lever

    (b) solenoid

    (c) hydraulic pilot

    (d) servo

Chapter 8

2. The purpose of a cylinder cushion is to decelerate the piston at the end of its stroke.

4. Given a 20 gpm pump, a load of 1000 pounds, and a 1 in. diameter rod in a 2 in. diameter piston, the responses are as follows:

(a)
$$p = \frac{F}{A}$$

$$p = \frac{1000}{3.1416}$$

$$p = 318.3 \text{ psi}$$

$$\text{area of piston} = \frac{\pi D^2}{4}$$

$$= 0.7854 D^2$$

$$= 0.7854(2)^2$$

$$= 3.1416 \text{ in.}^2$$

(b) velocity $= \dfrac{0.3208 \times 20}{A}$

$$v = \frac{0.3208 \times 20}{2.1416}$$

$$v = 2.04 \text{ ft/s for extension stroke}$$

(c) pressure on retraction is

$$p = \frac{F}{A}$$

$$p = \frac{1000}{2.9453}$$

$$p = 339.5 \text{ psi}$$

area of piston minus rod area

$$A = \frac{0.7854(1)^2}{4} \quad \text{area of rod} = 0.1963$$

effective area $= 3.1416 - 0.1963 = 2.9453$

(d) $v = \dfrac{0.3208 \times 20}{2.9453}$

$$= 2.17 \text{ ft/s for retraction stroke}$$

6. The main advantages of gear motors are simple design, lower costs, and ease of maintenance.

8. Given a hydraulic motor with a volumetric displacement of 2.31 in.^{3}/rev and a pressure rating of 3000 psi, connected to a 15 gpm pump, the following computations are provided

(a) speed

$$rpm = \frac{gpm \times 231}{displacement \ (in.^3/rev)}$$

$$rpm = \frac{15 \times 231}{2.31}$$

$$rpm = 1500$$

(b) $torque = \dfrac{pressure \times displacement \ (in.^3/rev)}{2\pi}$

$$T = \frac{pV_D}{6.28}$$

$$T = \frac{3000 \times 2.31}{6.28}$$

$$T = 1103.5 \ lb \cdot in.$$

(c) horsepower

$$HP = \frac{T \times rpm}{63,025} = \frac{1103.5 \times 1500}{63,025} = 26.26$$

10. Circuit requirements would be better served by a hydraulic piston motor than a gear motor where higher motor efficiency was needed, higher pressure was required, or maintenance services were available for hydraulic piston motors. Available space is also a consideration.

Chapter 9

2. An open-loop circuit is one in which there is no feedback. Control depends upon the characteristics of the components used and their interaction in the circuit. A closed-loop circuit uses feedback to produce a self-adjusting or self-regulating system.
4. A counterbalance valve is practical in a circuit where large forces must be prevented from dropping or restraining an overriding force such as a platen of a large press.
6. See Figure 9.13. Other arrangements are possible. See also Wolansky et. al. *Fundamentals of Fluid Power*, page 247 for other design. Cylinders or motors cannot be precisely synchronized over a large number of cycles of operation under varying load conditions using open-loop controls.

Chapter 10

2. A digital property device is either "on" or "off."
4. This is partially described in the text. Draw your own diagrams to follow the sequence.

Chapter 11

2. 5590 ft^3
4. 35.8 ft^3

INDEX